# Hemodynamic Monitoring in the ICU

Raphael Giraud • Karim Bendjelid

# Hemodynamic Monitoring in the ICU

 Springer

Raphael Giraud
Intensive Care Service
Geneva University Hospital
Geneva
Switzerland

Karim Bendjelid
Intensive Care Service
Geneva University Hospital
Geneva
Switzerland

ISBN 978-3-319-29429-2          ISBN 978-3-319-29430-8   (eBook)
DOI 10.1007/978-3-319-29430-8

Library of Congress Control Number: 2016939126

Printed on acid-free paper

This Springer imprint is published by Springer Nature
The registered company is Springer International Publishing AG Switzerland

# Preface

The benefit of any hemodynamic monitoring technique is to provide reliable and reproducible information on the cardiocirculatory status of a patient in shock. The collected values will enable the intensivist to understand the hemodynamic conditions of the patient and to make more informed treatment decisions to optimize the hemodynamic status and improve the prognosis of the patient. Hemodynamic monitoring is required to assess systemic and regional tissue perfusion as the correction of circulatory instability and tissue hypoperfusion is essential to prevent the occurrence of multiple organ failure. Resuscitation is characterized by a very close temporal relationship between monitoring, decision-making, and treatment. Indeed, making prompt, appropriate management and diagnostic and therapeutic decisions in cases of hemodynamic instability reduces mortality of critically ill patients [1].

To make treatment decisions, the intensivist has an arsenal of monitoring devices. However, before using a device, it is imperative that the intensivist has sound knowledge of the pathophysiology of shock states to identify the parameters that he/she wants to monitor. Therefore, knowing the different hemodynamic monitoring parameters is important [2]. For instance, it is now established and accepted by clinicians that fluid responsiveness is determined by monitoring dynamic indices, not static indices [3]. In addition, it is becoming increasingly common for mechanically ventilated critically ill patients to not be curarized. This condition requires clinicians to adapt their practice and to use other tactics to help assess the volume status of their patients. This is particularly true with the passive leg raising maneuver [4]. However, the issue at hand is the overall choice of monitoring technique. In the 1970s, the only advanced hemodynamic monitoring option was the pulmonary artery catheter. The use of the pulmonary artery catheter (PAC) has been challenged in recent years, and there has been debate regarding its impact on patient survival. Conflicting results have been published [5], though the widely varying conclusions are due to patient selection, incomplete information, and differences in specific treatment protocols (or lack thereof) [6]. In light of this and by following the developments in the industry, clinicians are now in favor of using less-invasive monitoring techniques. During the past few years, different techniques have been made commercially available. These devices are diverse in concept, design, and functionality but have more or less been shown to be reliable in clinical practice. Moreover, relative to the PAC, these devices are more easily handled, which could lead to their adoption and early application for use in large populations of at-risk patients or in patients with hemodynamic instability. However, for everyday use in clinical practice, the diversity of minimally invasive hemodynamic monitoring requires knowledge of a number of different techniques, their operating concepts, their settings, and their respective clinical validity.

In the first part of the present book, we present the hemodynamic monitoring parameters available to the clinician and their pathophysiological importance. For instance, blood pressure is the basic parameter, but measuring the arterial tone is sometime also necessary [7]. Additionally, measuring the intravascular pressure [8], the cardiac output, and their derived parameters is essential to determine and manage a balance between oxygen supply and consumption [9]. In this regard, we review techniques for cardiac output measurements based on pulmonary thermodilution [10], transpulmonary thermodilution [11, 12], echocardiography [13, 14], and Doppler techniques [15]. We discuss the techniques based on calibrated and non-calibrated pulse contour analysis [16] and their limitations. Finally, we discuss the dynamic indices of fluid responsiveness and their clinical applications and issues [17–22].

Geneva, Switzerland                                                    Raphael Giraud
                                                                       Karim Bendjelid

## References

1. Rivers E, Nguyen B, Havstad S, Ressler J, Muzzin A, Knoblich B et al (2001) Early goal-directed therapy in the treatment of severe sepsis and septic shock. N Engl J Med [Clinical Trial Randomized Controlled Trial Research Support, Non-U.S. Gov't] 345(19):1368–1377
2. Bendjelid K, Romand JA (2003) Fluid responsiveness in mechanically ventilated patients: a review of indices used in intensive care. Intensive Care Med 29(3): 352–360
3. Siegenthaler N, Giraud R, Saxer T, Courvoisier DS, Romand JA, Bendjelid K (2014) Haemodynamic monitoring in the intensive care unit: results from a web-based swiss survey. Biomed Res Int 2014:129593
4. Monnet X, Rienzo M, Osman D, Anguel N, Richard C, Pinsky MR et al (2006) Passive leg raising predicts fluid responsiveness in the critically ill. Crit Care Med 34(5): 1402–1407
5. Harvey S, Young D, Brampton W, Cooper AB, Doig G, Sibbald W et al (2006) Pulmonary artery catheters for adult patients in intensive care. Cochrane Database Syst Rev [Meta-Analysis Review] (3):CD003408
6. Takala J (2006) The pulmonary artery catheter: the tool versus treatments based on the tool. Crit Care 10(4):162
7. Chemla D (2006) Factors which may influence mean arterial pressure measurement. Can J Anaesth J Can Anesth 53(4):421–422
8. Rajaram SS, Desai NK, Kalra A, Gajera M, Cavanaugh SK, Brampton W et al (2013) Pulmonary artery catheters for adult patients in intensive care. Cochrane Database Syst Rev [Meta-Analysis Research Support, Non-U.S. Gov't Review] 2:CD003408
9. Creamer JE, Edwards JD, Nightingale P (1990) Hemodynamic and oxygen transport variables in cardiogenic shock secondary to acute myocardial infarction, and response to treatment. Am J Cardiol 65(20):1297–1300
10. Yelderman ML, Ramsay MA, Quinn MD, Paulsen AW, McKown RC, Gillman PH (1992) Continuous thermodilution cardiac output measurement in intensive care unit patients. J Cardiothorac Vasc Anesth 6(3):270–274

11. Giraud R, Siegenthaler N, Bendjelid K (2011) Transpulmonary thermodilution assessments: precise measurements require a precise procedure. Crit Care 15(5):195

12. Monnet X, Persichini R, Ktari M, Jozwiak M, Richard C, Teboul JL (2011) Precision of the transpulmonary thermodilution measurements. Crit Care [Clinical Trial] 15(4):R204

13. De Backer D (2014) Ultrasonic evaluation of the heart. Curr Opin Crit Care 20(3): 309–314

14. Vieillard-Baron A, Slama M, Cholley B, Janvier G, Vignon P (2008) Echocardiography in the intensive care unit: from evolution to revolution? Intensive Care Med [Review] 34(2):243–249

15. Monnet X, Rienzo M, Osman D, Anguel N, Richard C, Pinsky MR et al (2005) Esophageal Doppler monitoring predicts fluid responsiveness in critically ill ventilated patients. Intensive Care Med 31(9):1195–1201

16. Schloglhofer T, Gilly H, Schima H (2014) Semi-invasive measurement of cardiac output based on pulse contour: a review and analysis. Can J Anaesth J Can Anesth 61(5):452–479

17. Bendjelid K, Suter PM, Romand JA (2004) The respiratory change in preejection period: a new method to predict fluid responsiveness. J Appl Physiol 96(1):337–342

18. Cannesson M, Besnard C, Durand PG, Bohe J, Jacques D (2005) Relation between respiratory variations in pulse oximetry plethysmographic waveform amplitude and arterial pulse pressure in ventilated patients. Crit Care 9(5):R562–R568

19. Feissel M, Michard F, Faller JP, Teboul JL (2004) The respiratory variation in inferior vena cava diameter as a guide to fluid therapy. Intensive Care Med 30(9):1834–1837

20. Michard F (2011) Stroke volume variation: from applied physiology to improved outcomes. Crit Care Med 39(2):402–403

21. Michard F, Chemla D, Richard C, Wysocki M, Pinsky MR, Lecarpentier Y et al (1999) Clinical use of respiratory changes in arterial pulse pressure to monitor the hemodynamic effects of PEEP. Am J Respir Crit Care Med 159(3):935–939

22. Vieillard-Baron A, Chergui K, Rabiller A, Peyrouset O, Page B, Beauchet A et al (2004) Superior vena caval collapsibility as a gauge of volume status in ventilated septic patients. Intensive Care Med [Clinical Trial] 30(9):1734–1739

# Contents

# Abbreviations

| | |
|---|---|
| ACF | Aorto-caval fistula |
| ACP | Acute cor pulmonale |
| AHP | Arterial hypertension |
| AP | Arterial pressure |
| ARDS | Acute respiratory distress syndrome |
| C | Arterial compliance |
| $CaCO_2$ | Arterial content in $CO_2$ |
| $CaO_2$ | Arterial content in $O_2$ |
| CCO | Continuous cardiac output |
| CO | Cardiac output |
| COPD | Chronic obstructive pneumonia |
| CP | Cuff pressure |
| CPA | Cardiopulmonary arrest |
| CVC | Central venous catheter |
| $CvCO_2$ | Venous content in $CO_2$ |
| $CvO_2$ | Venous content $O_2$ |
| CVP | Central venous pressure |
| DBP | Diastolic blood pressure |
| $DO_2$ | $O_2$ delivery |
| dP/dtmax | The maximum rate of left ventricular pressure rise |
| DSt | Down slope time |
| DT | Deceleration time |
| dZ/dt | Maximal thoracic impedance variation |
| ECG | Electrocardiogram |
| ECMO | Extracorporeal membrane oxygenation |
| ET | Ejection time |
| ETVD | Right ventricular ejection time |
| EVLW | Extravascular lung water |
| $FeCO_2$ | Expired fraction in $CO_2$ |
| $FeN_2$ | Expired fraction in $N_2$ |
| $FeO_2$ | Expired fraction in $O_2$ |
| $FiCO_2$ | Inspired fraction in $CO_2$ |
| $FiN_2$ | Inspired fraction in $N_2$ |
| $FiO_2$ | Inspired fraction in $O_2$ |
| FS | Fractional shortening |
| GEDV | Global end-diastolic volume |
| GEF | Global ejection fraction |

| | |
|---|---|
| Hb | Hemoglobin |
| HR | Heart rate |
| ICT | Isovolumetric contraction time |
| ITBV | Intrathoracic blood volume |
| ITTV | Intrathoracic thermal volume |
| IVC | Inferior vena cava |
| IVRT | Isovolumetric relaxation time |
| LA | Left atrium |
| LAP | Left atrial pressure |
| LV | Left ventricle |
| LVEDD | Left ventricular end-diastolic diameter |
| LVEDP | Left ventricular end-diastolic pressure |
| LVEF | Left ventricular ejection fraction |
| LVESD | Left ventricular end-systolic diameter |
| LVET | Left ventricular ejection time |
| MAP | Mean arterial pressure |
| MSP | Mean systemic pressure |
| MTt | Mean transit time |
| PAC | Pulmonary arterial catheter |
| $PaCO_2$ | Arterial partial pressure in $CO_2$ |
| PAH | Pulmonary arterial hypertension |
| $PA_{lv}$ | Alveolar pressure |
| $PaO_2$ | Arterial partial pressure in $O_2$ |
| PAOP | Pulmonary arterial occluded pressure |
| PAP | Pulmonary arterial pressure |
| PBV | Pulmonary blood volume |
| Pcap | Pulmonary capillary pressure |
| PEEP | Positive end-expiratory pressure |
| PEP | Pre-ejection period |
| PP | Pulse pressure |
| PTV | Pulmonary thermal volume |
| $PvCO_2$ | Venous partial pressure in $CO_2$ |
| PVI | Plethysmography variability index |
| $PvO_2$ | Venous partial pressure in $O_2$ |
| PvP | Pulmonary venous pressure |
| PVPI | Pulmonary vascular permeability index |
| PWS | Pulse wave speed |
| PWV | Pulse wave velocity |
| RA | Right atrium |
| RAP | Right atrial pressure |
| RV | Right ventricle |
| RVEF | Right ventricular ejection fraction |
| RVP | Right ventricular pressure |
| $SaO_2$ | Arterial saturation in $O_2$ |
| SBP | Systolic blood pressure |
| $ScvO_2$ | Central venous saturation in $O_2$ |
| SPV | Systolic blood pressure variation |
| $StO_2$ | Tissue saturation in $O_2$ |

| | |
|---|---|
| SV | Stroke volume |
| SVC | Superior vena cava |
| $SvO_2$ | Mixed venous saturation in $O_2$ |
| SVR | Systemic vascular resistance |
| SVV | Stroke volume variation |
| TAPSE | Tricuspid annular plane systolic excursion |
| TEE | Transesophageal echocardiography |
| TPR | Total peripheral resistance |
| TPTD | Transpulmonary thermodilution |
| TR | Tricuspid regurgitation |
| TTE | Transthoracic echocardiography |
| $VCO_2$ | $CO_2$ production |
| VES | Ventricular extra systole |
| $VO_2$ | $O_2$ consumption |
| VR | Venous return |
| VSD | Ventricular septal defect |
| VTI | Velocity time integral |
| $Z_0$ | Basal impedance |
| $\Delta IVC$ | Inferior vena cava respiratory variation |
| $\Delta PEP$ | Respiratory variation in pre-ejection period |
| $\Delta Pleth$ | Respiratory variation in plethysmography |
| $\Delta PP$ | Pulse pressure variation |
| $\Delta SVC$ | Superior vena cava respiratory variation |

# Introduction

The management of a critically ill patient in shock requires the monitoring of physiological parameters of the cardiovascular system. The goal is to detect physiological anomalies and to provide the clinician with information to make diagnoses and define treatment strategies. However, if the use of monitoring techniques is not evaluated or validated, the type of monitoring used and the high degree of invasiveness of such techniques may present issues.

Acute circulatory failure is a common condition in intensive care and is clinically significant, affecting the prognosis of patients. Cardiogenic shock is related principally to myocardial infarction, with a 30–50 % mortality rate. The mortality rate of septic shock patients is 20–40 %, and the management of these patients requires the use of hemodynamic monitoring. Clinicians must be alerted by a low cardiac output, which is difficult to detect based solely on clinical arguments. Data from the literature showing that clinicians were unable to identify more than 50 % of states of shock based only on clinical observations [1]. Moreover, the persistence of low CO can lead to multiple organ failure.

From a physiological point of view, cardiovascular monitoring can be divided into two categories: monitoring of the macrocirculation and of the microcirculation.

For each monitoring technique, knowledge of the method, measurement technique, and their characteristics is essential. As is the case for all assays, each technique should have an associated accuracy, which corresponds to the approximation of a measure compared with a reference sample, and a precision that matches the variability of several measurements. For instance, the measurement of cardiac output by the classical right heart thermodilution has a coefficient of variation of approximately 12 %. However, the reliability of the present measurements can be altered by tricuspid insufficiency, intracardiac shunts, or congenital heart disease.

Over the past decade, the results of randomized studies showed no improved mortality rates for ICU patients fitted with pulmonary artery catheters. Patient monitoring was greatly reduced, and less invasive techniques were used. During this period, echocardiography became an increasingly popular tool for static measurements for cardiologists in intensive care units. This was also an opportunity for the industry to develop a range of "noninvasive" measuring devices that allowed the clinician to obtain parameters such as pulse heart that were typically traditionally assessed with "invasive" techniques.

However, as no monitoring method has been shown to be responsible for morbidity, it seems unreasonable to disregard the assistance provided by such tools for both diagnosing and monitoring patients in critical situations. Unfortunately, since clinical and paraclinical parameters are commonly insufficient for identifying the nature of cardiovascular disorders in complex situations such as circulatory failure, important informations may be missed if substantial further monitoring is not conducted.

Currently, in the intensive care, there are no ideal hemodynamic monitoring methods that can provide accurate, reproducible, reliable, and noninvasive information on all parameters of the cardiovascular system. The ideal tool should provide information to the clinician to determine appropriate adjustments to resuscitation treatments such as volume expansion and inotrope or vasopressor use, which would ultimately correct circulatory disorders and improve patient health.

In the absence of this ideal tool, a multitude of cardiovascular exploration techniques are available for the intensivist.

## Reference

1. Chioléro R, Revelly JP (2003) Concept de monitorage hémodynamique en soins intensifs. Rev Med Suisse 538(2462)

# Blood Pressure

During shock, a very common clinical situation in the ICU, the measurement of systemic blood pressure is an essential component of the diagnosis, severity, therapeutic management, and patient monitoring. This measurement is one of the first variables monitored by clinicians. Blood pressure is a controlled variable of the cardiovascular system, and hypotension indicates a significant disruption of homeostasis.

Blood pressure values help to provide quantitative informations. In fact, these numbers are compared with the threshold values that define a shock state, permitting a positive diagnosis. However, blood pressure must be interpreted based on the comorbidities in each individual patient (e.g., age, hypertension, heart failure, diabetes, and standard treatments). The blood pressure figures have predictive values and represent a therapeutic target in the treatment of shock. Blood pressure also provides qualitative informations. Indeed, with the measurement of cardiac output and CVP, it can be used to calculate the peripheral vascular resistance, allowing a differential diagnosis of shock from all these determinants.

Blood pressure comprises four components: systolic blood pressure, diastolic blood pressure, mean blood pressure, and pulse pressure. The combined study of these four elements is used to define a hemodynamic profile. In addition, the shape of the blood pressure curve can aid in the diagnoses of certain diseases. For instance, when studying the respiratory variations in pulse pressure ($\Delta PP$) in patients with a regular heartbeat who are placed on controlled mechanical ventilation and have a tidal volume greater than 8 ml/kg, the fluid responsiveness can be predicted if the PPV is greater than 13 % [1]. These patients, called "responders," are able to increase their cardiac output by over 15 % after intravenous fluid infusion.

## 1.1 Blood Pressure Measurement

The measurement of blood pressure can be carried out invasively or noninvasively.

### 1.1.1 Noninvasive Measurement

Intra-arterial pressure is indirectly measured when an "occlusive" methods measure intermittent local flow changes caused by the pressure against an inflated cuff. These changes include auscultatory changes, as detected by a sphygmomanometer or by the transmission of small oscillations to the cuff via the "oscillometric" method. "Nonocclusive" methods measure the continuous intra-arterial pressure transmitted in the perivascular region of a slightly flattened artery through the "tonometry" method or the contra-pressure necessary to maintain a constant volume of the digital arteries by photoplethysmography.

© Springer International Publishing Switzerland 2016
R. Giraud, K. Bendjelid, *Hemodynamic Monitoring in the ICU*, DOI 10.1007/978-3-319-29430-8_1

The noninvasive reference measurement of blood pressure was once performed using a mercury sphygmomanometer with the auscultation method (Figs. 1.1 and 1.2).

Mercury sphygmomanometers were phased out due to environmental concerns. Currently, a cuff placed preferably on the brachial artery is inflated to a pressure above the systolic pressure. The cuff is then deflated slowly. The appearance [2] and disappearance [3] of Korotkoff sounds (turbulent flow) correspond to the SBP and the DBP, respectively. The auscultatory method is preferred over the palpation method, which only measures the SBP (Fig. 1.3).

This method is difficult to use in the ICU, especially during emergency situations. The fact that this method is manual means that it does not allow for the automatic monitoring of blood pressure. In addition, SBP measurement is dependent on the local blood flow of a pulsating turbulent flow, which is responsible for the sounds heard in phase I. It is therefore highly dependent on the distal vasomotor tone. Therefore, auscultation is difficult or impossible to measure, especially during severe hypotension or shock state. Finally, increased arterial stiffness, as is observed in the elderly or in patients suffering from atherosclerosis, may also cause the brachial artery to be less compressible and can alter the transmission of Korotkoff sounds. This leads to the underestimation of SBP measured by a sphygmomanometer and the overestimation of DBP.

The "oscillometric" method measures small oscillations of the backpressure induced in a vessel when an occlusive cuff deflates according to a commercially protected algorithm. This corresponds to the transmission of the arterial pulsation when the flow is restored. Gradually, as the cuff deflates, these oscillations pass through a maximum and then decrease and disappear. Devices that use this method measure only the MAP (and calculate SBP and DBP) as the

**Fig. 1.1** Mercury column sphygmomanometer

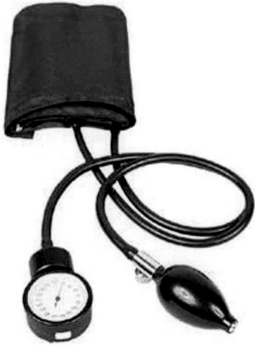

**Fig. 1.2** Blood pressure cuff

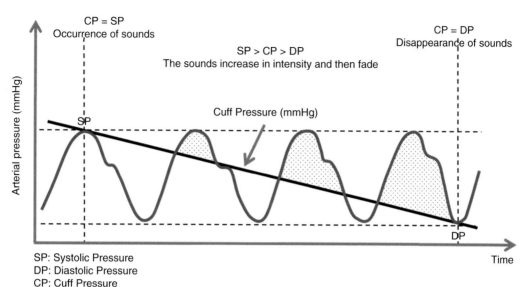

CP > SP
Absence of sound

CP = SP
Occurrence of sounds

CP = DP
Disappearance of sounds

SP > CP > DP
The sounds increase in intensity and then fade

Cuff Pressure (mmHg)

Arterial pressure (mmHg)

SP

DP

Time

SP: Systolic Pressure
DP: Diastolic Pressure
CP: Cuff Pressure

**Fig. 1.3** Principle of noninvasive blood pressure measurement with the use of a sphygmomanometer and stethoscope. First, the cuff pressure (*CP*) is increased higher than systolic pressure to block blood flow in the arm. Then, the cuff pressure is gradually decreased until the blood pressure is at a sufficient level to enable blood to pass through the artery. This is the systolic pressure. The cuff pressure is continually decreased to a value at which there is no obstacle in the arterial flow (laminar flow), even in diastole. This is the diastolic pressure. This diagram shows the relationship between blood pressure, cuff pressure, and sounds of the artery

contra-pressure corresponding to the maximal oscillations.

The method of "digital photoplethysmography" measures cyclical fluctuations in blood flow that enters and leaves the finger, preferably the index finger, and provides values corresponding to the finger blood volume. This method is based on the transmission of light through the finger. A diode emits infrared light to measure the digital volume. It is connected to a system that assigns this volume to a pressure required to maintain the digital volume and a constant arterial volume. This technique is known as the "volume-clamp" and allows the continuous monitoring of beat-to-beat blood pressure [4].

Tonometry is a method that has been used for decades by ophthalmologists to measure intraocular pressure. Recently, it has been developed to measure pressure in superficial arteries, preferably the radial artery. This method involves applying a slight pressure with a pressure transducer formed by a piezoresistive crystal on the skin over the radial artery. By overcoming the extramural

pressure, this method allows the intramural pressure transmitted to the sensor to be continuously measured. The calibration of the radial signal is performed assuming that the MAP and DBP are identical between the brachial and radial arteries. The central aortic pressure curve can then be reconstructed using a transfer function and validated in a large population of patients. This method could more precisely quantify the pulsatile component of the afterload of the left ventricle at the central aortic level [5]. Although validated in stable patients under general anesthesia [3], the relevance of this technique in patients in shock [6–8] remains to be demonstrated.

### 1.1.2 Invasive Blood Pressure Measurement

Invasive AP measurement is preferred in all cases where the reliability of the noninvasive measurement is questionable, for both its bad precision (e.g., arrhythmias and extremes of hypo- or hypertension, as technical difficulties in obese trauma) and the lack of continuous measurements when sudden changes are expected, especially when patients are receiving vasoactive drugs, positive inotropic, and/or intravenous antihypertensive treatments. Invasive blood pressure measurements avoid distortions from over- or underestimations of AP (mainly at the expense of SBP and DBP values), which are dependent on the characteristics of the hydraulic system, representing the "weakest" point of the measurement chain. Currently, preassembled systems with an electrical pressure transducer are available for clinical use. These disposable blood pressure transducer systems deliver vital accuracy in invasive blood pressure measurement with lowest possibility of zero drift. A careful purge of the circuit is necessary to avoid signal interference due to bubbles in the circuit. Thus, MAP is a precise parameter that is directly measured from the area under the blood pressure curve over time and is used in cases of arrhythmia, with measurement errors of generally less than 2 %. In elective situations, in a patient with an SBP >80 mmHg, the preferred puncture site is the radial artery (Fig. 1.4).

Using an Allen test is recommended to assess the presence and condition of the collateral network [9]. In the case of shock or emergency conditions, the femoral artery is often preferred. Brachial and dorsalis pedis arteries may be used as alternatives. Teflon or polyurethane catheters with a maximum diameter of five French for the femoral artery or three French for the radial artery are recommended. Using a purge system for a continuous flow of 2 ml/h, with the possibility of achieving an intermittent manual purge, is also recommended. The addition of heparin showed no benefit [10]. The most feared complication of arterial catheterization is arterial thrombosis [11]. The prevention of thrombosis is dependent on the choice of material, implementation of the Allen test, size of the catheter, and duration of catheterization. Another severe complication is infection [12]. The application of aseptic measures equivalent to those required for the establishment and use of central venous lines ensures its prevention [13].

Arterial catheterization is the reference and gold standard technique for beat-by-beat arterial blood pressure measurements. Significant intraindividual differences were reported compared with noninvasive techniques [14,15]. However, in the context of emergencies and particularly in prehospital care, noninvasive methods have been shown to be the only usable option despite their relative unreliability, especially in the case of hypotension [16]. Arterial catheterization provides beat-by-beat information on blood pressure values (i.e., SBP, DBP, MAP, and PP) and also enables visualization

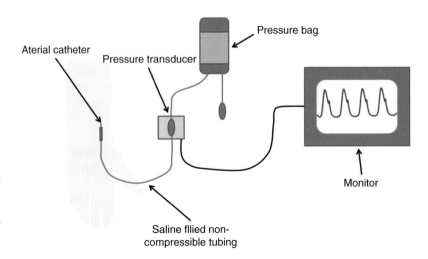

**Fig. 1.4** Arterial catheterization measuring principle. The arterial catheter is connected via a pressure tube to a pressure sensor, which is connected to a monitor

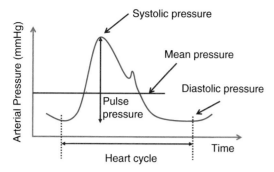

**Fig. 1.5** Blood pressure curve

of the blood pressure curve (Fig. 1.5). These two elements are the basis of hemodynamic monitoring practices in the ICU [17].

## 1.2    Mean Arterial Pressure

### 1.2.1    Definition, Calculation, and Normal Values

MAP is the pressure that ensures blood flow in case where we don't take in account the pulsatility [18]. Thus, it corresponds to the pressure that provides organ perfusion, except that of the left ventricle, which is perfused by the left coronary artery principally in diastole. MAP is calculated by measuring the area under the blood pressure curve and dividing that value by the cardiac cycle duration (over time). ICU monitoring devices typically average blood pressure values over several seconds.

When a sphygmomanometer and stethoscope are used to measure arterial blood pressure, MAP can be estimated by the formula:

$$MAP = \frac{2 \times DBP + 1 \times SBP}{3}$$

If the oscillometric method is used, the measured value is really the MAP and the systolic and diastolic components are mathematically derivated by protected algorithms (patents). From a physiological point of view, whatever the localization of the measurement (i.e., brachial, radial, femoral, and carotid arteries), MAP is considered constant [5,19,20]. MAP, like all intravascular

pressures, is related to atmospheric pressure (760 mmHg). The zero reference is made at the level of the heart.

### 1.2.2    Pressure, Flow Resistance

Blood flow is driven by the difference in total energy between two points. Although pressure is normally considered as the driving force for blood flow, in reality, it is the total energy that drives flow between two points. In this regard, the study of non-pulsatile flow at a constant rate is based on resistance [21]. When balanced, MAP is the pressure that theoretically provides the same cardiac output in continuous mode (i.e., not pulsed) according to the relationship:

$$(MAP - MSP) - (RAP - MSP) = SVR \times CO$$

which means that $MAP - RAP = SVR \times CO$, where MSP is the mean systemic filling pressure, i.e., the theoretical pressure present throughout the circulatory system when the blood flow is zero, RAP is the right atrial pressure, SVR is the systemic vascular resistance, and CO is the cardiac output. The present driving pressure gradient is analogous with the potential $U$ difference across a circuit comprising a resistor $R$ and a current $I$ and governed by Ohm's law (i.e., $U = R \times I$).

In shock, treatments that are administered to enhance MAP increase the cardiac output (i.e., volume replacement, positive inotropic) or vascular resistance (vasopressor). Mean *systemic* filling pressure (MSP), though often misconcepted with mean circulatory filing pressure (MCFP) and often comparable in value, is different. MSP represents the pressure generated by elastic recoil in the systemic circulation during a no-flow state. MSP is not measurable in clinical practice but can be observed during the death of a patient, seconds after cardiac arrest.

Hence,

$$(MAP - RAP) = SVR \times CO$$

Thus, MAP is determined by CO, SVR, and RAP according to the relationship:

$$MAP = (CO \times SVR) + RAP$$

Often, only the systemic vascular resistances (also called total peripheral resistances) are calculated:

$$SVR = \frac{MAP}{CO}$$

and RAP is neglected when the present value is low (<5 mmHg). However, the SVR has no straightforward physiological significance either at rest or during dynamic maneuvers [22].

### 1.2.3  Blood Viscosity, Resistance Vessels

SVR is not a measured variable but rather is calculated from the measured values of MAP, RAP, and CO. However, SVR is not simply a theoretical value. The quantity that characterizes the difficulty of a fluid to flow is its viscosity. As part of the laminar system in which inertial forces are neglected, Poiseuille's law can be applied to the systemic circulation to characterize the present difficulty:

$$SVR = \frac{8\eta L}{\pi r^4}$$

where $\eta$ is the blood viscosity, $L$ is the length of the functional vascular network, and $r$ is the radius of the functional systemic vessels. The level of resistance in vessels, i.e., the contraction or relaxation of smooth muscle cells (linked to mechanical stimuli or mediated by endothelial function), is small relative to decreases or increases in the functional radius $r$. Systemic vascular resistances are inversely proportional to the fourth power of the functional radius $r$; this results in a significant increase or decrease in the resistance. Along the arterial tree, the largest average pressure drop is observed at the arterioles and capillaries (resistance vessels). The aorta and its branches and some smaller arteries (in particular, brachial and radial arteries) have very low resistances. The measured MAP values in the upper and lower limbs are accurate representations of the central aortic MAP [23].

### 1.2.4  Information Provided by MAP and Changes in MAP

MAP is closely related to CO, resistivity, and mean systemic pressure by the equation: $MAP = (CO \times SVR) + RAP$.
 Notably,

- SVR is not measured but is calculated from MAP, RAP, and CO:

$$SVR = (MAP - RAP)/CO$$

- In patients with shock, especially in the case of right heart failure, tamponade, or fluid infusion maneuvers, RAP may play a major role in this equation.

The self-regulation of MAP is a key element of the cardiovascular system [18]. In physiology, a sharp decrease in MAP is normally offset by sympathetic stimulation, leading to reflex tachycardia, increased stroke volume (due to a positive inotropic effect and increase in preload related to venoconstriction) and systemic arterial vasoconstriction. In patients experiencing septic shock or vasoplegia, these compensatory mechanisms are often outdated or defective. Therefore, a fall in the MAP may be caused by a decrease in the cardiac output that is insufficiently offset by reflex sympathetic vasoconstriction or a disproportionate decrease in SVR due to vasodilatation. Accordingly, it is essential for these patients to be monitored for cardiac output to precisely determine the properties responsible for decreases in blood pressure [22].

MAP is often regarded as a related and controlled variable of the cardiovascular system with several determinants. It is determined by various regulatory mechanisms, including baroreflex. This reflex is initiated by mechanoreceptors sensitive to deformation. These receptors called "baroreceptors" are located in the walls of large systemic arteries. High-pressure baroreceptors are located in the carotid sinus, aortic arch, and right atrium; low-pressure baroreceptors are located in the pulmonary vessels. The neurons constituting these baroreceptors have relays in the nucleus of the solitary tract located in the medulla. The pulses from the carotid baroreceptors are not detected if the MAP is below 60 mmHg. They gradually appear with increasing blood pressure to a maximum of 180 mmHg. When activating the baroreceptors, the integration of signals at the nucleus of the solitary tract leads to both the inhibition of the sympathetic neurons located in the rostral ventrolateral medulla and the excitement of the cardiac vagal neurons located in the ambiguous nucleus and the dorsal vagal nucleus. The vasomotor center manages the efferent signal to the heart and blood vessels and thus influences the vascular-cardiac coupling. The baroreflex response is opposite that of blood pressure [24].

MAP values in the large arteries are often stable; accordingly, the MAP is considered the perfusion pressure in most vital organs. When the MAP falls below the lower limit of the autoregulation plateau, regional blood flow becomes linearly dependent on the MAP [25]. The lower limit of the self-regulating plate is 60–70 mmHg. These limits vary with the cardiovascular history of each patient, the considered organ, the pathology, the metabolic activity, and the use of vasodilators.

The autoregulation of organ blood flow, which is the tendency for organ blood flow to remain constant despite changes in the arterial perfusion pressure, is a ubiquitous phenomenon. Four mechanisms of autoregulation, myogenic, metabolic, tissue pressure, and tubuloglomerular feedback, have been recognized as potentially important, whereas a fifth possible mechanism, local neural control, has been noted but given little credence. Over the years, substantial evidences have been obtained in support of the metabolic, myogenic, and tubuloglomerular feedback mechanisms of blood flow autoregulation. The relative contribution of the metabolic and myogenic mechanisms varies considerably among vascular beds as well as among few tissues; usually, a single mechanism is apparently responsible for autoregulation. For organs in which both mechanisms are present, we do not have a good

understanding of their relative importance. Moreover, the contribution of each mechanism may vary according to the metabolic activity of the tissue and experimental conditions of the study. The metabolic or flow-dependent mechanism of blood flow autoregulation appears to depend on tissue oxygen levels, probably acting through alterations in tissue metabolism. However, a direct effect of oxygen on the resistance vessels cannot be excluded. The heterogeneous nature of tissue $PO_2$ distribution suggests that areas of low oxygen tension may act as main position of flow regulation by producing vasodilator substances as oxygen delivery falls. The identity of specific chemical mediators produced in such hypothesized areas remains to be determined [25].

Hypotension is defined when the MAP <60 mmHg [26]. In patients with a history of hypertension, a decrease in the MAP of more than 40 mmHg is considered hypotension, even though the pressure is above 60 mmHg. However, there is no minimal MAP that ensures an adequate perfusion of all organs since the critical value of the MAP is different for each organ. Therefore, there are only recommendations, especially in septic shock, in which the target minimum MAP is 65 mmHg to prevent organ hypoperfusion [27]. These recommendations are based on clinical studies that have shown that MAP >65 mmHg do not improve organ perfusion or tissue oxygenation [27]. However, the elderly or hypertensive patients do require higher MAP levels.

## 1.3   The Pulse Pressure

### 1.3.1   Definition of Capacitance Vessels

The proximal portions of the arterial system (i.e., the aorta and its first divisions) are rich in elastin and are thus "elastic" [21]. Indeed, these vessels have the ability to cushion cardiac ejections by absorbing part of the systolic ejection volume and eventually restoring the volume during diastole (Windkessel phenomenon). This provides the distal edge network with continuous blood flow. The primary utility of this operation is consuming less energy than if it had to contain the arterial vessels without completely absorbing the stroke volume.

### 1.3.2 Pulse Wave Velocity and Concept of Reflected Waves

The impact of heart beat and stroke volume on the arterial vasculature is perceptible as a pulse wave. The pulse wave observed in arterial compliance-resistance depends on both the incident wave and the reflected wave. The pulse wave velocity is inversely proportional to the arterial compliance. The pressure increase is dependent on the compliance ($C$) so that $C = dV/dP$, as shown in a few studies [28,29]. The blood pressure level is directly related to the physical properties of the arterial tree. The interactions between blood pressure (SBP, DBP, MAP, and PP), the function of the left ventricle (LV), and peripheral resistance have long been unclear. The Windkessel model [30] attempted to represent these interactions. However, this model was limited. Studies have shown that another model using pulse wave propagation was more suitable.

The Windkessel model was originated in 1899 [30]. This was a physical model used to express the changes in blood pressure, cardiac output, peripheral vascular resistance, and arterial compliance over time:

$$dP(t)/dt = \frac{I(t)}{C} - \frac{P(t)}{RC}$$

where $P(t)$ is the change in blood pressure, $I(t)$ is the flow that corresponds (with reference to an electrical system) to the intensity of the current and thus to the CO, $dP$ is the potential difference of the circuit, $R$ is the SVR, and $C$ is the arterial compliance. This model was inspired by an electrical system with a generator (i.e., the heart), a resistor (i.e., peripheral resistance), and a capacitor (i.e., arterial compliance). The blood pressure corresponds to the circuit voltage (potential difference), and the cardiac output corresponds to the current intensity. During systole, the compliant arterial system absorbs some of the blood volume ejected by the LV. The stroke volume is then returned during diastole so that the pulse wave is dampened along the arterial tree and the overall flow is constant. This model accredits the arterial system a blood pressure damping function and a blood flow transfer function (Fig. 1.6).

However, this model has a number of limitations [31]. First, it does not explain the

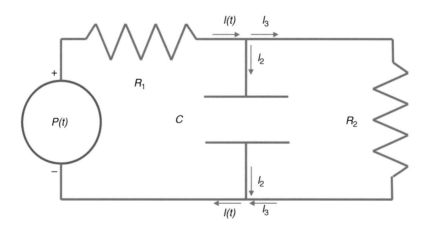

**Fig. 1.6** Three-element Windkessel model. $R$ represents resistors (where $R_1$ is the resistance due to the aortic valves and $R_2$ is the peripheral resistance), $P(t)$ is the blood pressure variation, $C$ is the vascular compliance, $I$ is the current intensity, $I(t)$ is the flow from the pump as a function of time, $I_2$ is the current in the middle branch of the circuit, and $I_3$ is the current in the right branch of the circuit

pressure amplification phenomenon. Second, the propagation of the pulse wave velocity (PWV) is not taken into account. As the PWV is dependent on the compliance of the considered arterial segment, there is a direct relationship between the arterial compliance and PWV, which can be described according to Bramwell and Hill's equation [32]:

$$\text{PWV} = \sqrt{\frac{V\ dP}{\rho dV}}$$

where $\rho$ corresponds to the density of the blood, $dP/dV$ is the compliance, and $V$ is the volume.

### 1.3.3   The Current Model

Based on these findings and aimed to explain the integrative function of cardiac and arterial functions, a more realistic model was formulated that incorporates the propagation of the pulse wave throughout the arterial tree [31]. Indeed, the interaction between the LV, peripheral resistance, and arterial compliance involves studying the shape and the propagation velocity of the wave. The current model can better explain the changes observed along the arterial tree as a pressure amplification phenomenon, especially in aging, hypertension, and heart failure or during the use of vasodilators. The pulse wave generated by the LV is transferred to the elastic proximal aorta. The incident wave is reflected at arterial bifurcations so that the wave reflections are added to the incident wave (merging phenomenon). The reflected wave returns earlier or later, depending on the reflected distance and the pulse wave velocity (which depends from the compliance of the vascular system) [29,31,33]. During episodes of vasoconstriction, the reflection sites are close to the central point of emission. There is not only one single reflected wave. Several small waves occur and coalesce into a "reverse" wave [31]. Thus, the pulse wave varies and is dependent on all phenomena affecting vascular compliance, i.e., the patient age, the blood pressure measurement site, and the patient status (i.e., hypertension and heart failure).

In young subjects, the central pulse wave has an early systolic peak and low amplitude (Fig. 1.7a). The reflection waves rebound (arrow) before another wave is propagated and continues after diastole. The reflection waves gradually decay over time.

In the elderly, the central pulse wave (Fig. 1.7b) has larger and later peak systolic amplitudes and later and earlier return waves (due to decreased vessel compliance). A superposition of the systolic peak and return waves is responsible for the peak amplitude. The lack of influence of the return wave after closing the sigmoid is responsible for a higher diastolic decay. The pressure (i.e., the difference between the peak and the first shoulder) gradually increases with age (high value of pulse pressure).

A difference exists between the distal (femoral) and central pressures (aorta). Indeed, the central artery systolic pressure is lower than the pressure of the peripheral arteries. The change of the wall structure (central arteries are elastic and more muscular) accounts for the decreasing distensibility of the arteries from the center to the periphery. Increasing the PWV and reflection sites accounts for the earlier peak incident pressure and returning waves. The measured peripheral blood pressure is greater than the brachial pressure, which is itself higher than the central pressure. The brachial pressure is not a good substitute measurement for central pressure, especially in young patients.

In patients suffering from hypertension, there is a change in the shape of the pulse wave, as observed in the elderly and explained by the same phenomena. There is a greater arterial stiffness and an increased PWV due to an increased peripheral resistance with a greater early return waves in the cycle (Fig. 1.7c)

Finally, in patients with a low CO, there are two peaks in the pulse wave. The return waves occur after closure of the sigmoid cusps due to a shortening of the left ventricular ejection time. Therefore, the first peak corresponds to the systolic peak of the left ventricle, and the second peak corresponds to a diastolic peak generated by the return waves. Premature sigmoid cusp closures in these patients may be explained by this

**Fig. 1.7** Pulse wave.
(**a**) Pulse wave in the
young. (**b**) Pulse wave in
the elderly. (**c**) Pulse
pressure increase. The
pressure (difference
between the peak and the
first shoulder) gradually
increases with age

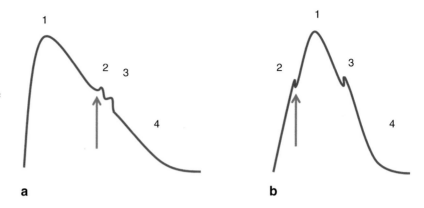

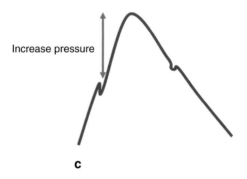

phenomenon involving the return wave. During
the administration of vasodilators, the amplitude
and the early return waves increase the ejection
time and the cardiac output [31,34].

### 1.3.4   The Aortic Pulse Pressure

The PP is the difference between the systolic and
diastolic blood pressure. Its value, measured at a
determined location in the arterial system, is a
function of the ejected blood volume and the
artery elasticity [21]. PP determination considers
periodic flows involving concepts that are similar
to impedance and circuits. In the simplest model,
in the proximal part of the aorta, the system is
similar to that of a two-element air tank: the two-
element Windkessel model, which comprises a
capacitive element corresponding to the total
arterial compliance $C$ that is added in parallel to
the SVR. The decrease in blood pressure during
diastole has a monoexponential character that

can be expressed by its arterial time constant
(Tau) as follows:

$$Tau = SVR \times C$$

This model can be applied to the most proximal
part of the ascending aorta. Thus, it is possible to
estimate the arterial compliance with this simplified approach based on the two-element
Windkessel model [21]:

$$C = \frac{SV}{Aortic\ PP}$$

More complex models involving the proximal
aortic elasticity such as the "characteristic impedance" and reflection wave models exist [35].

### 1.3.5   Peripheral Pulse Pressure

Peripheral PP is not closely related to the
MAP and is even less related to the SVR. It
reflects the pulsatile component of blood

pressure and is primarily dependent on arterial stiffness (1/compliance which equals elastance) and SV. A study performed in ICU patients showed that peripheral PP was primarily due to SV and the total arterial compliance *C*, as measured by arterial tonometry [36]. Thus, it is possible to interpret the PP respiratory variations in patients under mechanical positive pressure ventilation to estimate the preload dependency [1,37]. Increased arterial stiffness in elderly patients results, for a same stroke volume, to an increase in the PP associated with higher SBP and lower DBP. Therefore, a low PP observed in an elderly patient is likely due to an impaired SV. Indeed, arterial stiffness in these patients is assumed to be high [38]. Finally, the continuous monitoring of PP in the ICU can indirectly follow SV variations [39].

## 1.4 Diastolic Blood Pressure

It is commonly accepted by clinicians that vascular tone is the main determinant of DBP. A decrease in the vascular tone (e.g., consecutive to a vasoplegic state) will result in a decrease in DBP. In the absence of severe aortic regurgitation, DBP is dependent on the diastolic period and the period of decreasing blood pressure, which is in turn dependent on the compliance and resistance of blood according to the equation:

$$Tau = SVR \times C$$

With each heartbeat, an increased diastolic time will result in a lower DBP, and conversely, a decrease in diastolic time will result in a higher DBP. Similarly, a shorter decay time (Tau) will result in a decrease in DBP, and conversely, DBP will be higher if Tau increases. In the case of vasoplegia or arterial compliance disturbance associated with increased arterial stiffness, Tau will be shorter. Therefore, for a given level of MAP and/or SVR, an increased arterial stiffness is associated with a lower DBP; additionally, increased arterial stiffness decreases DBP twice as much unless SBP is increased.

## 1.5 Systolic Blood Pressure

The total arterial compliance and characteristics of the left ventricular ejection fraction are regarded by clinicians as the main determinants of SBP. However, as for DBP, the physiological and pathological realities cannot be so easily described [20]. In fact, for a given level of MAP or SVR, a high SBP will be observed in the case of increased arterial stiffness, and the increased SBP observed will be twice as large as the concomitant decrease in DBP. Similarly, for a given level of MAP or SVR, decreased arterial stiffness in relation to improved compliance will be associated with a lower SBP, and SBP decreases twice as much as the concomitant increase in DBP. Acute alterations of the elastic properties of large arteries causing acute changes in arterial compliance have been studied, especially during shock.

## References

1. Michard F, Boussat S, Chemla D, Anguel N, Mercat A, Lecarpentier Y et al (2000) Relation between respiratory changes in arterial pulse pressure and fluid responsiveness in septic patients with acute circulatory failure. Am J Respir Crit Care Med 162(1):134–138
2. Chioléro R, Revelly JP (2003) Concept de monitorage hémodynamique en soins intensifs. Rev Med Suisse 538(2462)
3. Vos JJ, Poterman M, Mooyaart EA, Weening M, Struys MM, Scheeren TW et al (2014) Comparison of continuous non-invasive finger arterial pressure monitoring with conventional intermittent automated arm arterial pressure measurement in patients under general anaesthesia. Br J Anaesth 113(1):67–74
4. Philippe EG, Hebert JL, Coirault C, Zamani K, Lecarpentier, Y, Chemla D (1998) A comparison between systolic aortic root pressure and finger blood pressure. Chest 113(6):1466–1474
5. O'Rourke MF, Adji A (2005) An updated clinical primer on large artery mechanics: implications of pulse waveform analysis and arterial tonometry. Curr Opin Cardiol 20(4):275–281
6. Ameloot K, Van De Vijver K, Van Regenmortel N, De Laet I, Schoonheydt K, Dits H et al (2014) Validation study of Nexfin(R) continuous non-invasive blood pressure monitoring in critically ill adult patients. Minerva Anestesiol 80:1294–1301
7. Fischer MO, Avram R, Carjaliu I, Massetti M, Gerard JL, Hanouz JL et al (2012) Non-invasive continuous

arterial pressure and cardiac index monitoring with Nexfin after cardiac surgery. Br J Anaesth 109(4): 514–521

8. Ruiz-Rodriguez JC, Ruiz-Sanmartin A, Ribas V, Caballero J, Garcia-Roche A, Riera J et al (2013) Innovative continuous non-invasive cuffless blood pressure monitoring based on photoplethysmography technology. Intensive Care Med 39(9):1618–1625, Research Support, Non-U.S. Gov't Validation Studies

9. Hildick-Smith D (2006) Use of the Allen's test and transradial catheterization. J Am Coll Cardiol 48(6):1287, author reply 8

10. Robertson-Malt S, Malt GN, Farquhar V, Greer W (2014) Heparin versus normal saline for patency of arterial lines. Cochrane Database Syst Rev 5:CD007364

11. Scheer B, Perel A, Pfeiffer UJ (2002) Clinical review: complications and risk factors of peripheral arterial catheters used for haemodynamic monitoring in anaesthesia and intensive care medicine. Crit Care 6(3):199–204

12. O'Horo JC, Maki DG, Krupp AE, Safdar N (2014) Arterial catheters as a source of bloodstream infection: a systematic review and meta-analysis. Crit Care Med 42(6):1334–1339

13. Zingg W, Cartier V, Inan C, Touveneau S, Theriault M, Gayet-Ageron A et al (2014) Hospital-wide multi-disciplinary, multimodal intervention programme to reduce central venous catheter-associated bloodstream infection. PLoS One 9(4):e93898

14. Jones DW, Appel LJ, Sheps SG, Roccella EJ, Lenfant C (2003) Measuring blood pressure accurately: new and persistent challenges. JAMA 289(8):1027–1030

15. Pickering TG, Hall JE, Appel LJ, Falkner BE, Graves J, Hill MN et al (2005) Recommendations for blood pressure measurement in humans and experimental animals: Part 1: blood pressure measurement in humans: a statement for professionals from the Subcommittee of Professional and Public Education of the American Heart Association Council on High Blood Pressure Research. Hypertension 45(1):142–161

16. Cohn JN (1967) Blood pressure measurement in shock. Mechanism of inaccuracy in auscultatory and palpatory methods. JAMA 199(13):118–122

17. Michard F (2005) Changes in arterial pressure during mechanical ventilation. Anesthesiology 103(2):419–428, quiz 49-5

18. Chemla D (2006) Factors which may influence mean arterial pressure measurement. Can J Anaesth = J Can Anesth 53(4):421–422

19. Pauca AL, O'Rourke MF, Kon ND (2001) Prospective evaluation of a method for estimating ascending aortic pressure from the radial artery pressure waveform. Hypertension 38(4):932–937

20. Franklin SS, Gustin W, Wong ND, Larson MG, Weber MA, Kannel WB et al (1997) Hemodynamic patterns of age-related changes in blood pressure. The Framingham Heart Study. Circulation 96(1):308–315

21. Chemla D, Hebert JL, Coirault C, Zamani K, Suard I, Colin P et al (1998) Total arterial compliance estimated by stroke volume-to-aortic pulse pressure ratio in humans. Am J Physiol 274(2 Pt 2):H500–H505

22. Badeer HS, Hicks JW (1994) Pitfalls in the assessment of vascular resistance. Cardiology 85(1):23–27

23. McEniery CM, Yasmin, Hall IR, Qasem A, Wilkinson IB, Cockcroft JR (2005) Normal vascular aging: differential effects on wave reflection and aortic pulse wave velocity: the Anglo-Cardiff Collaborative Trial (ACCT). J Am Coll Cardiol 46(9):1753–1760

24. Hainsworth R (1990) Non-invasive investigations of cardiovascular reflexes in humans. Clin Sci 78(5): 437–443

25. Johnson PC (1986) Autoregulation of blood flow. Circ Res 59(5):483–495

26. Practice parameters for hemodynamic support of sepsis in adult patients in sepsis. Task Force of the American College of Critical Care Medicine, Society of Critical Care Medicine (1999) Crit Care Med 27(3):639–660

27. Asfar P, Meziani F, Hamel JF, Grelon F, Megarbane B, Anguel N et al (2014) High versus low blood-pressure target in patients with septic shock. N Engl J Med 370(17):1583–1593, Comparative Study Multicenter Study Randomized Controlled Trial Research Support, Non-US Gov't

28. Ong KT, Delerme S, Pannier B, Safar ME, Benetos A, Laurent S et al (2011) Aortic stiffness is reduced beyond blood pressure lowering by short-term and long-term antihypertensive treatment: a meta-analysis of individual data in 294 patients. J Hypertens 29(6): 1034–1042

29. O'Rourke M (1991) Arterial compliance and wave reflection. Arch Mal Coeur Vaiss 84(Spec No 3): 45–48

30. Frank O (1899) Die Grundform des arteriellen Pulses. Z Biol 37:483–526

31. Laurent S, Cockcroft J, Van Bortel L, Boutouyrie P, Giannattasio C, Hayoz D et al (2006) Expert consensus document on arterial stiffness: methodological issues and clinical applications. Eur Heart J 27(21):2588–2605

32. Westenberg JJ, van Poelgeest EP, Steendijk P, Grotenhuis HB, Jukema JW, de Roos A (2012) Bramwell-Hill modeling for local aortic pulse wave velocity estimation: a validation study with velocity-encoded cardiovascular magnetic resonance and invasive pressure assessment. J Cardiovasc Magn Reson: Off J Soc Cardiovasc Magn Reson 14:2, Research Support, Non-US Gov't Validation Studies

33. Merillon JP, Motte G, Masquet C, Azancot I, Guiomard A, Gourgon R (1982) Relationship between physical properties of the arterial system and left ventricular performance in the course of aging and arterial hypertension. Eur Heart J 3(Suppl A):95–102

34. O'Rourke MF (2009) Time domain analysis of the arterial pulse in clinical medicine. Med Biol Eng Comput 47(2):119–129

35. Dart AM, Kingwell BA (2001) Pulse pressure – a review of mechanisms and clinical relevance. J Am Coll Cardiol 37(4):975–984

36. Lamia B, Teboul JL, Monnet X, Osman D, Maizel J, Richard C et al (2007) Contribution of arterial stiffness and stroke volume to peripheral pulse pressure in ICU patients: an arterial tonometry study. Intensive Care Med 33(11):1931–1937

37. Michard F, Chemla D, Richard C, Wysocki M, Pinsky MR, Lecarpentier Y et al (1999) Clinical use of respiratory changes in arterial pulse pressure to monitor the hemodynamic effects of PEEP. Am J Respir Crit Care Med 159(3):935–939

38. Kelly R, Hayward C, Avolio A, O'Rourke M (1989) Noninvasive determination of age-related changes in the human arterial pulse. Circulation 80(6): 1652–1659

39. Marik PE, Cavallazzi R, Vasu T, Hirani A (2009) Dynamic changes in arterial waveform derived variables and fluid responsiveness in mechanically ventilated patients: a systematic review of the literature. Crit Care Med 37(9):2642–2647, Meta-Analysis Review

# Monitoring of Cardiac Output and Its Derivatives

## 2.1 Method of Measuring Cardiac Output with the Pulmonary Artery Catheter

The classical pulmonary artery catheter, also called the Swan-Ganz catheter, is a hemodynamic monitoring tool still in use in the ICU. Over the last 15 years, less invasive techniques have been developed, resulting in reduced use of this catheter. In addition, a meta-analysis on the influence of this catheter on the survival of ICU patients showed no benefit in terms of mortality rates [1, 2]. However, it is important to understand what information can be obtained using the present catheter. Here, the authors present the data that can be obtained through the pulmonary artery catheter and the resulting calculations: intravascular pressure, cardiac output (by the thermodilution technique), and mixed venous oxygen saturation levels.

### 2.1.1 Dilution Techniques of an Indicator

The following equation summarizes the measurement principle:

$$\Phi = m / \sum_{o}^{t} C_e \, dt$$

where $\Phi$ is the flow rate, $m$ is the mass of the indicator, $C_e$ is the concentration of the indicator to the output (assuming that the concentration of

the indicator to the input is zero), and $\sum_{o}^{t} C_e \, dt$ is the area under the curve corresponding to the concentration over time (no recirculation). This is the Stewart-Hamilton principle.

### 2.1.2 Thermodilution

#### 2.1.2.1 Intermittent Measurement Using the "Bolus" Technique

This technique relies on the conservation of thermal energy. The method involves injecting a cold bolus of fluid at the inlet of the core. To correctly estimate the flow, there must be no loss of indicator between the injection site and the detection site. The indicator must also be completely mixed with the blood. Finally, the temperature variation between the injectate and the base temperature must be easily detected. Therefore, testing both the sensitivity and the precision of the thermistor (to detect temperature differences of 0.01 °C) is necessary before using a pulmonary artery catheter. This step, however, is often overlooked in practice.

The main equation is as follows [3]:

$$\Phi_b = \frac{q_i \, S_i \, Q_i \left( T_b - T_i \right)}{q_b \, S_b \sum_{t_1}^{t_2} \Delta T_b \left( t \right) dt}$$

where $\Phi_b$ is the blood flow, $T_b$ and $T_i$ are the temperatures of the blood and the indicator,

© Springer International Publishing Switzerland 2016
R. Giraud, K. Bendjelid, *Hemodynamic Monitoring in the ICU*, DOI 10.1007/978-3-319-29430-8_2

respectively, $Q_i$ is the injected volume of the cold bolus, $q$ and $S$ represent the specific gravity and the heat of the injectate and the base temperature of the blood, and $t_1$ and $t_2$ are the injection time and the end time of integration, respectively, when the cold liquid passes in front of the detector. Due to heat transfer at the catheter wall, the temperature of the injected bolus corresponding to the intra- and extrathoracic catheter is not uniform either during or at the end of the injection. This can lead to a prolonged thermodilution curve in the descending portion, leading to a phenomenon called "recirculation." The descending part of the curve is often extrapolated from measurement programs according to a monoexponential pattern, which is acceptable only if the blood flow is stable and there is no basis for thermal line deflection. Therefore, the equation can be simplified as follows:

$$\Phi_b = \frac{K_1 K_2 \Phi_1 (T_b - T_1)}{\sum_o^t \Delta T_b(t)\mathrm{d}t + A}$$

where $K_1 = \Phi_i S_i / q_b S_b$; $K_2$ is a constant that depends on the dead space of the catheter, temperature variations, and the injection rate; $t$ is the final timepoint; and $A$ is the area under the curve. A calculation constant (CC) is inserted either manually or automatically in the calculation program and is dependent on the type of catheter (e.g., size, length of intravascular route, thermal conductivity of the wall), the type of fluid (e.g., 0.9 % NaCl, $K_1 = 1.10$; 5 % glucose, $K_1 = 1.08$), the fluid bolus temperature (e.g., cold or room temperature), and the injected volume. No significant differences exist between the various commercially available computer programs [4].

In several clinical studies, thermodilution was compared using the Fick method and the indocyanine green dilution method with very satisfactory results. However, to use this technique for monitoring the most critically ill patients, accounting for the measurement variability is essential. A 22 % variation in the CO is required to establish a clinically significant change in the CO with the injection of a single bolus; with the successive injection of a second bolus, a variation of 15 % is required [5, 6].

Measurement errors are dependent on:

**The Method**

If the CC is too low, resulting in CO underestimation, then a correction of the calculation constant is possible using the following formula:

$$\text{Correct CO} = \left( CC_{correct} / CC_{incorrect} \right) \text{incorrect CO}$$

The injection is usually performed through the proximal lumen of the pulmonary artery. However, it can also be performed at the right ventricle using another central catheter or at the lateral channel of the introducer. The results from either method are contradictory and are dependent on the injected volume [7]. In fact, the CO measurement precision appears to decrease when the injection volume decreases from 10 (cold) to 5 ml (ambient temperature <25 °C) for CO ranging from 4.7 to 7.7 l/min. In this case, the variability increases due to the decrease in the signal to noise ratio [7]. In contrast, no differences were detected for CO <4.7 l/min. Finally, for a CO > 7.7 l/min, the accuracy decreases, and the variability significantly increases when the injection volume varies from 10 (cool) to 5 ml (room temperature). Therefore, if the CO is normal or low, a bolus of 5–10 ml of cold liquid at room temperature may be used interchangeably. However, if the CO is high and in circumstances where the signal to noise ratio is particularly low (e.g., when the ambient temperature is above 25 °C such as in the case of hypothermia or significant respiratory changes), it is essential to use a cold liquid bolus of 10 ml. Notably, the remainder of the solution is chilled for 45–60 min at ambient temperature. In addition, the injection should be completed within 30 s following preparation of the solution to avoid warming the cold liquid 64. The use of an automatic injection method is likely to avoid warming the liquid, but its use is not satisfactory in clinical practice. Using a "closed circuit" technique in combination with a temperature measurement system at the injection site, as proposed by the CO-set® system, would reduce the risk of bacterial contamination. This would be required every 24 h. The use of this technique remains limited because of its cost.

## The Clinical Situation

Underestimation of the CO due to recirculation of the indicator is possible in cases of tricuspid or pulmonary insufficiency and the existence of a left-right shunt. However, the CO may also be overestimated in case of an underestimation of the area under the curve (loss of the indicator as in the right-left shunt). In addition, CO variations may occur in the case of chills, cough, and heart rhythm disorder induced by cold injections. These parameters can affect the measurement. Errors of 15–50 % in the CO measurement by thermodilution are possible if the base temperature changes by more than 0.05 °C, which can occur randomly or cyclically or after a cardiopulmonary bypass [8]. Most calculation programs used do not account for these variations. Therefore, the thermodilution curve must be clearly examined to eliminate CO measurement values corresponding to aberrant curves. The dispersion of CO values is increased during mechanical ventilation [9]. The values vary depending on the time of the bolus injection [10] and the respiratory frequency [11]. Therefore, although performing the injection at a fixed time during the respiratory cycle reduces the variability of the measurement, it does not provide a good CO estimation [10]. Moreover, the increase in respiratory rate decreases the variability, yet it can lead to hemodynamic changes and $O_2$ diffusion, causing difficulty in interpreting the CO [11]. Increasing the number of measurements made for random breath cycles ($\geq$4 injections) or performing measurements at regular injection intervals (4 injections) is better to reduce the risks ($R$) of more than 10 % deviations from the reference value (4 injections at random, $R \leq 20$ %; 4 injections at regular intervals, $R=0$ %) [3].

In summary, the method for CO measurement using the classical bolus thermodilution technique is reliable. However, the user must meet a number of technical requirements, which are easily overlooked given the trivialization of the method. Finally, if the CO varies over time [12], a variation of 12–15 % between two CO determinations is considered clinically significant (three measurements per determination) [6].

### 2.1.2.2 Thermodilution Curves

The typical appearance of a normal curve has a bottom segment due to the rapid injection of injectate followed by a smooth curve and a slightly prolonged return descent to the baseline. Because this thermodilution curve represents temperature variation from the highest to the lowest temperature (blood temperature decrease) followed by a return to a higher temperature, the "true" curve is in a negative direction. The area under the curve (Riemann integral) is inversely proportional to the CO (Fig. 2.1). In the case of low CO, the return to baseline is prolonged, resulting in a greater area under the curve. At a high cardiac output, the coldest liquid bolus travels faster through the heart: the temperature then returns quickly to the base value, resulting in a smaller area under the curve (Fig. 2.2).

### 2.1.2.3 Continuous Measurement Method

A continuous CO measurement technique based on the principle of thermodilution has been developed. The indicator used to measure the CO is a quantity of heat on the order of 7.5 W and is randomly delivered in the right ventricle (mixing chamber) by means of a heating filament. This increases the signal to noise ratio at the right chamber, thereby improving the accuracy of CO measurements. The statistical properties of the input and output signals are taken

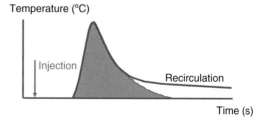

**Fig. 2.1** Temperature curve as a function of time for CO calculation by pulmonary artery thermodilution. The calculation of the area under the thermodilution curve using the Stewart-Hamilton formula permits determination of the CO

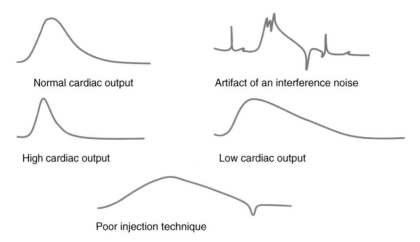

**Fig. 2.2** Examples of different pathological and/or incorrect thermodilution curves obtained by pulmonary artery thermodilution

into account for the stochastic analysis curves. This is not the case with thermodilution, for which only the instantaneous values of the temperature signal are taken into account. A pseudo-random binary code "injected" into the right ventricle introduces a quantity of heat between 10 and 15 W. Each random sequence corresponds to 15 states that are either open or closed (Fig. 2.3a). Temperature variations are detected at the output of the system, i.e., in practice, at the level of the pulmonary artery (Fig. 2.3b). A cross-correlation is then performed between the input and output areas (Fig. 2.3c). The result of this analysis is then used to construct a thermodilution curve, and the CO is evaluated by measuring the area under the curve.

The formula is as follows [13]:

$$F = \frac{P(60)/4.180}{(2N/N+1)\rho c \sum_{K=0}^{N-1} \Phi_{\Delta T}(t)(k\Delta t)}$$

where $\rho$ is the density in g/L (1.05 for blood); $\Delta T$ is the temperature variation; $F$ is the flow in l/min; $c$ is the specific heat, i.e., a binary pseudorandom code (1 or −1); $\Delta t$ is the duration of the status (open or closed); $N$ is the period of the pseudorandom sequence; $P$ is the heat of the filament (in watts); and $\sum_{K=0}^{N-1} \Phi_{\Delta T}(t)(k\ t)$ is the area under the curve obtained by cross-correlation between the input and output signals.

A Swan-Ganz catheter is connected to a computing device (e.g., Vigilance™, Edwards Life Sciences, Irvine, California). The "open" state duration is variable but is present for nearly half the time. The calculation phase is automatic, without any intervention necessary by the user. No calibration is required. After initialization and the start of CO calculations, an initial value appears after a variable time and is updated every 30–60 s. The display reflects the CO values from 3 to 6 min earlier. An unfiltered display mode is possible with the Vigilance™ calculation program. The filament, located between 14 and 25 cm from the distal end of the catheter, should be placed as close as possible to the pulmonary valve to allow a homogeneous mixture of the thermal indicator to minimize any CO calculation errors.

The first comparison study of CO obtained by continuous thermodilution utilized a conventional technique wherein a liquid bolus showed an average difference of 0.02 l/min between the two techniques, with bias from −1.03 to 1.07 l/min [13]. These results are also confirmed by comparison with the thermodilution method (bolus), the indocyanine green method, and the Fick method [14–19]. However, the dispersion of the values around the average is dependent on the study. A tendency to increase the CO bias exists when the CO is high [16, 18, 19]. If these results are explained by an increase of the variability of the measured CO values according to the conventional liquid bolus method [20], an increase in the

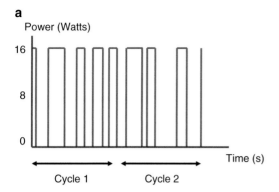

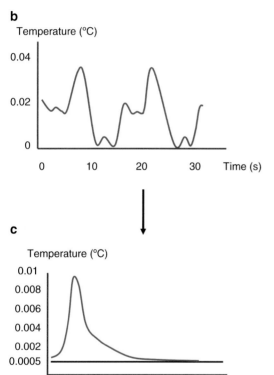

**Fig. 2.3** The CO thermodilution continuous measurement method. (**a**) Input signal corresponding to two cycles of a binary pseudorandom code; (**b**) output signal corresponding to the change in temperature at the level of the distal pulmonary artery following the "injection" of heat at the right ventricular level according to the binary code described in (**a**); (**c**) cross-correlation between a and b to calculate CO by measuring the area under the curve

difficulty. Therefore, this method offers several advantages over the conventional bolus method. It is independent of the operator and eliminates the errors associated with variations in volume, temperature, and injection speed. It reduces variations in the CO due to ventilation (because the automatic reported average considers the entire respiratory cycle) and the risk of bacterial contamination related to bolus injections. This method does allow CO evolution to be analyzed over time.

The complex shape of the right ventricle, which is wound around the left ventricle, makes the measurement of its surface and volume difficult via echocardiography. Determining the ejection fraction is too complex to practically use this variable for hemodynamic monitoring. However, analysis of the temperature curve recorded during a cardiac cycle permits the fraction of the remaining volume in the right ventricle after ejection to be derived. If there is no arrhythmia or significant tricuspid regurgitation, the monitor can then deduct the residual temperature and thus the RV ejection fraction [21]. The RV diastolic volume can also be calculated by dividing the stroke volume by the ejection fraction. Compared with a standard system, this method comprises a triple proximal injectate port and a rapid-response thermistor for acquisition every 52 ms. As a low-pressure pump, the RV is particularly sensitive to afterload, and obviously, the systolic performance decreases when the pulmonary vascular resistance increases, the normal value of the RVEF being approximately 40–50 % [22].

The ventricular end-diastolic volume (EDV) is a better indicator than the filling pressure of preload. The correlation between the RVEDV and the CO is much higher than that with the CVP or the PAOP [23]. The correlation remains high in positive-pressure-ventilated patients subjected to PEEP, whereas the CVP and PAOP have no valid correlation in these situations. A normal RVEDV is in the range from 60 to 100 ml/m$^2$. A value greater than 120 ml/m$^2$ is associated with RV dilatation [21].

Critics of this type of sophisticated and expensive measurement claim that the RV ejection fraction is highly dependent on its afterload and does not actually measure contractility. Furthermore, coupling between the RVEDV and the CO is

response time in measurement of the CO by the continuous thermodilution method must be considered [16]. No incidents have been published regarding the long-term use of this catheter. Although the size of this catheter is larger than the standard catheter, its insertion poses no particular related to the mathematical formulation of the calculation: the RVEDV is calculated from the ejection fraction and the SV (i.e., two shared variables are compared, but one is used to calculate the other [mathematical coupling]).

## 2.2    Transpulmonary Thermodilution

### 2.2.1    Measurement of Cardiac Output by Transpulmonary Thermodilution

The insertion of a pulmonary artery catheter was found to be too cardiac invasive and irrelevant in terms of the survival rate in critically ill patients. In this regard, transpulmonary thermodilution surpassed (in terms of the number of catheters placed) pulmonary artery catheter use. This technique requires a central venous line located in the intrathoracic portion and an arterial catheter with a thermistor located in a large artery (i.e., the femoral or brachial artery) with a tip assumed to be not so far from the thorax. The technique involves injecting a cold bolus (<8 °C) in the venous catheter and collecting the temperature curve via the thermistor on the arterial catheter (Fig. 2.4).

The CO is calculated according to the modified Stewart-Hamilton equation (Fig. 2.5).

Transpulmonary thermodilution has been validated against pulmonary arterial thermodilution [24–27] and the Fick method [28]. It is a reliable and reproducible method for measuring the cardiac output in both adults and children [27, 29]. Three bolus injection of cool liquid (often NaCl 0.9 %) are required to obtain a cardiac output value. Additionally, transpulmonary thermodilution is less influenced by breathing than the Swan-Ganz catheter. The cold indicator injected by the central venous line is successively diluted in the right heart chambers, the pulmonary circulation, the left heart chambers, and the descending aorta. Therefore, the CO is calculated using the Stewart-Hamilton equation, and the cold indicator distribution volumes are determined by mathematical

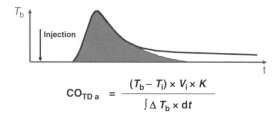

$$CO_{TD\,a} = \frac{(T_b - T_i) \times V_i \times K}{\int \Delta T_b \times dt}$$

**Fig. 2.5**  Measurement of the CO by the transpulmonary thermodilution technique and flow calculation by the Stewart-Hamilton equation. *Tb* temperature of blood, *Ti* temperature of injectate, *Vi* volume of injectate, $\int \Delta Tb^*dt$ integral of the area under the thermodilution curve, and *K* correction factor dependent on the weight, specific blood temperature, and injectate

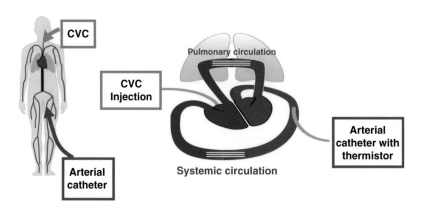

**Fig. 2.4**  Principle of the transpulmonary thermodilution technique with the injection of a cold fluid bolus via a central venous line located in the SVC region and temperature collection through an arterial catheter located in the femoral artery (Reprinted with permission from Pulsion® Medical System)

analysis of the thermodilution curve, i.e., the global end-diastolic volume (i.e., the amount of blood contained in the four cardiac chambers and the descending aorta) and the extravascular lung water volume.

## 2.2.2 Measurement of Global End-Diastolic Volume and Intrathoracic Blood Volume

### 2.2.2.1 Global End-Diastolic Volume

From mathematical analysis of the thermodilution curve, the mean transit time and the downslope time of the thermal indicator permit evaluation of the global end-diastolic volume (Fig. 2.6).

The distribution volume of the thermal indicator (Fig. 2.7) is obtained by multiplying the CO by the average transit time. This corresponds to the intrathoracic thermal volume, which comprises both the intrathoracic blood volume and the extravascular lung water volume (Fig. 2.8). The pulmonary thermal volume comprises both the pulmonary blood volume and the extravascular lung water volume. The pulmonary thermal volume is obtained by multiplying the CO by the downslope time, as the present chamber is assumed to be the larger volume located in the thorax, in comparison with heart chambers and

**Fig. 2.6** Measurement of the mean transit time (*MTt*) and downslope time (*DSt*) from the transpulmonary thermodilution curve and its logarithmic transformation (Reprinted with permission from Pulsion® Medical System)

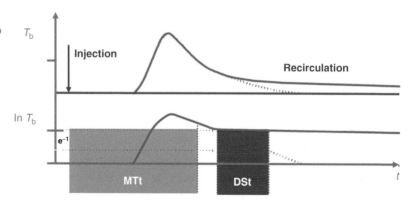

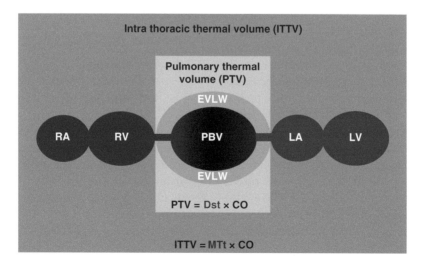

**Fig. 2.7** Schematic representation of the distribution of volumes of cold liquid bolus by the transpulmonary thermodilution technique. *MTt* is the average transit time, *RA* is the right atrium, *RV* is the right ventricle, *PBV* is the pulmonary blood volume, *LA* is the left atrium, *LV* is the left ventricle, *DSt* is the downslope time, *ITTV* is the intrathoracic thermal volume, and *EVLW* is the extravascular lung water (Reprinted with permission from Pulsion® Medical System)

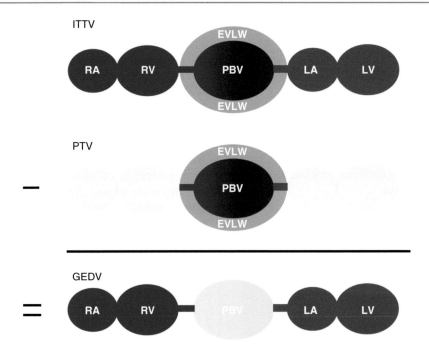

**Fig. 2.8** Principles for determining the global end-diastolic volume (*GEDV*) by the transpulmonary thermodilution technique. ITTV corresponds to the intrathoracic thermal volume, and EVLW corresponds to the extravascular lung water (Reprinted with permission from Pulsion® Medical System)

vessels [30, 31]. Therefore, the global end-diastolic volume is the difference between the intrathoracic thermal volume and the pulmonary thermal volume [32].

The global end-diastolic volume values are between 600 and 800 ml/m$^2$ [33, 34]. In reality, the GEDV is greater than the real heart blood volume. However, it is very dependent on the ventricular blood volume. Therefore, the GEDV is assumed to correspond to the cardiac preload index. Volume expansion will make the GEDV grow, and dobutamine will have no impact on its value. In addition, in the case of volume expansion, GEDV variations follow those of the left ventricular diastolic area and cardiac output. Therefore, increasing the GEDV follows the cardiac output without necessarily varying in the same direction, even though the two values are determined by the same thermodilution curve [33, 34]. In fact, there is no mathematical coupling between the cardiac output and the GEDV. Some diseases can cause abnormally high GEDV, as is the case for aortic

aneurysms or atrial dilation, wherein the GEDV is dependent on the blood volume between the cold liquid bolus injection site (central venous catheter) and the indicator detection site (femoral artery).

### 2.2.3   Calculation of the Intrathoracic Blood Volume

The intrathoracic blood volume (ITBV) is the sum of the GEDV and the pulmonary blood volume (PBV) (Fig. 2.9). The PBV is equal to approximately 20 % of the ITBV 180. From the formula, it is possible to estimate the intrathoracic blood volume:

The relationship between the ITBV and the GEDV is relatively strong and is not influenced by weight, size, cardiac output, vasoactive drugs, pulmonary hypertension, hypoxemia, or hypovolemia [35, 36]. Therefore, transpulmonary thermodilution can reliably estimate the ITBV from

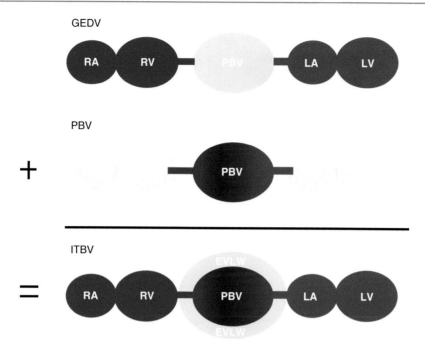

**Fig. 2.9** Evaluation principles of intrathoracic blood volume (*ITBV*) by transpulmonary thermodilution (Reprinted with permission from Pulsion® Medical System)

measurement of the GEDV. The ITBV is calculated as follows:

$$Intrathoracic\ blood\ volume = 1.25 * Global\ end\text{-}diastolic\ volume$$

(Fig. 2.10).

The ITBV was established by many studies as an indicator of the volume or preload [37, 38]. Its normal values are between 750 and 1,000 ml/m². Although the ITBV has no advantage compared with the GEDV in estimating blood volume, its measurement is essential for determining the extravascular lung water via transpulmonary thermodilution.

### 2.2.4  Measurement of Extravascular Lung Water

The extravascular lung water (EVLW) is calculated as the difference between the intrathoracic thermal volume and intrathoracic blood volume. Normal values of the EVLW are between 3 and 7 ml/kg. However, due to the limitations of the measurement technique, a value greater than 10 ml/kg is considered pathological in pulmonary edema. The EVLW shows a good correlation with the extravascular lung water measured by gravimetry in animal studies [39, 40] or by a double dilution method in humans [32, 35]. EVLW values of 35–40 ml/kg are observed in cardiogenic pulmonary edema or lesions [35]. The EVLW has been shown to be an independent prognostic index in critically ill patients. The greater the EVLW value, the more the prognosis appears to be affected [41].

Rarely, the EVLW may be useful in identifying patients in pulmonary edema when the clinical diagnosis of ARDS is difficult [42]. The EVLW has shown applicability in mechanical ventilation weaning tests [43] and eliminates the need to measure PAOP during spontaneous breathing trials [44].

Based on the ratio EVLW/PBV, it is possible to automatically calculate the pulmonary vascular permeability index (PVPI) (Fig. 2.11). For patients with acute lung injury and ARDS, PVPI will be high [45]. This index can be used to distinguish between inflammatory pulmonary edema (with increased pulmonary vascular permeability)

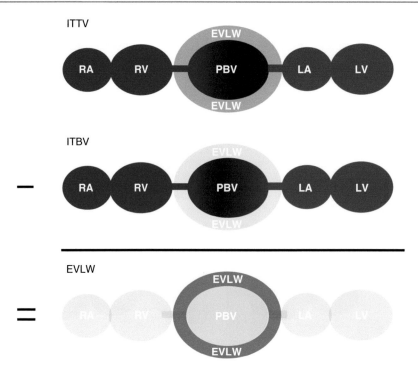

**Fig. 2.10** Principles of measurement of the extravascular lung water (*EVLW*) from the intrathoracic thermal volume (*ITTV*) and the intrathoracic blood volume (*ITBV*) (Reprinted with permission from Pulsion® Medical System)

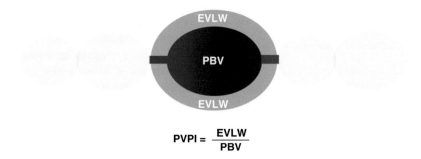

$$PVPI = \frac{EVLW}{PBV}$$

**Fig. 2.11** Assessment principles of the pulmonary vascular permeability index (*PVPI*) by the ratio of the extravascular lung water (*EVLW*) and the pulmonary blood volume (*PBV*) (Reprinted with permission from Pulsion® Medical System)

and hydrostatic pulmonary edema from a cardiogenic origin. When EVLW is greater than 12 ml/kg, PVPI values greater than 3 indicate an injury mechanism of pulmonary edema, with good specificity and sensitivity [46].

However, there are limitations to using the EVLW. Two clinical studies on patients with ARDS (diagnosed by conventional criteria) showed that the percentage of those with normal extravascular lung water was not negligible [47, 48]. This was explained by the occurrence of microthrombi in the pulmonary circulation in this pathology (increase in dead space). This caused a decrease in the dilution space of the cold indicator and, hence, lowered the EVLW values. In addition, in the case of an intracardiac shunt, recirculation of the cold indicator initially causes a significant elongation of the thermodilution curve. By calculating the EVLW from this curve and in the event of recirculation, the EVLW values are unusually high. The values are corrected when the shunt is closed [47].

### 2.2.5 Calculation of the Global Ejection Fraction

Hemodynamic parameters are more or less dependent on the ventricular preload and afterload conditions. The assessment of cardiac contractility is relatively difficult at the bedside. The ventricular ejection fraction is the most used clinical parameter for assessing right and left ventricle contractility and function [49]. The left ventricular ejection fraction is the ratio of the stroke volume to the ventricular end-diastolic volume. In addition, the ratio of the stroke volume to the global end-diastolic volume is called the global ejection fraction (GEF). This index can be considered the overall index of cardiac function. In determining the stroke volume by the ratio of the cardiac output on the heart rate to the global end-diastolic volume, transpulmonary thermodilution technique provides the global ejection fraction. The index can be used to reveal a right and/or left ventricular dysfunction [50]. However, it is impossible to determine which of the two ventricles fails when the global ejection fraction is low (less than 18–20 %). Therefore, it is necessary to refer to an echocardiography, which will help in understanding the mechanism of heart failure. Monitoring of the GEF can also be used to monitor the treatment [51].

### 2.2.6 Transpulmonary Thermodilution Allows for Calibration of the Contour Analysis Derived from the Arterial Pulse Wave

Some hemodynamic monitors continuously display the continuous cardiac output in real time using the pulse contour analysis of the arterial pulse wave from an arterial catheter. The reliability of this monitoring is based not only on the quality of the blood pressure signal but also on the initial calibration. Some monitors (e.g., LIDCO™, London, UK) use the lithium dilution method, whereas others use the transpulmonary thermodilution method to calibrate the system. Pulse contour analysis allows the continuous measurement of the CO and calculates an index of left ventricular contractility. The continuous CO is calculated as follows:

$$CO = SV \times HR$$

In the Pulsion® Medical System, SV is obtained by integration, systole by systole, of the area under the systolic part of the arterial pressure curve and is multiplied by a calibration factor (Fig. 2.12). The calibration factor is primarily dependent on the ventriculo-arterial coupling of the patient's cardiovascular system (arterial compliance, systemic vascular resistance, etc.). Measurement of the cardiac output by transpulmonary thermodilution is necessary for calibration especially during the administration of vasoactive drugs [24]. The left ventricular contractile performance is evaluated by measuring the pressure rise gradient ($dP_{max}$) during ventricular ejection. This is equal to the maximum instantaneous differential blood pressure ($dP/dt_{max}$). Finally, analysis of the variations of pulse contour of the arterial wave allows for measurement of the respiratory changes in the stroke volume (SVV) during a period of 7.5 s, which encompasses at least one complete respiratory cycle [52]. This is possible only in patients with controlled ventilation with positive pressure. An SVV greater than 9.5 % predicts an increase in the stroke volume after fluid expansion, suggesting the presence of relative hypovolemia [53].

Thus, the injection of a saline bolus by the central venous route permits the determination of the discontinuous transpulmonary thermodilution parameters described in this chapter and the automatic recalibration of the continuous measurement of cardiac output derived from pulse contour analysis.

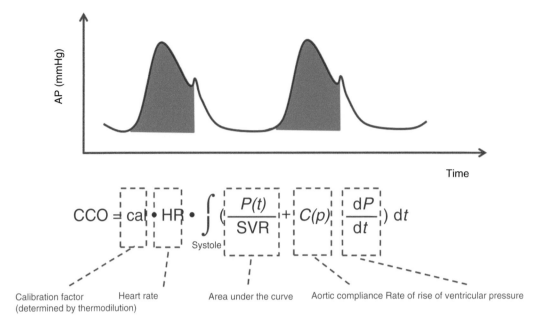

$$CCO = \mathrm{cal} \cdot \mathrm{HR} \cdot \int_{\mathrm{Systole}} \left( \frac{P(t)}{SVR} + C(p) \frac{dP}{dt} \right) dt$$

Calibration factor (determined by thermodilution)     Heart rate     Area under the curve     Aortic compliance   Rate of rise of ventricular pressure

**Fig. 2.12** Calculation of the continuous cardiac output (*CCO*) beat by beat by analyzing the pulse wave contour and calibrating by transpulmonary thermodilution by integrating the systolic part of the arterial pressure curve

## 2.3 Measurement of Cardiac Output Using a Chemical Indicator

Current methods use chemical indicator dilution (lithium chloride) to measure cardiac output, e.g., (LIDCO™, London, UK) [54]. With the present technique, pulse wave contour analysis does not use a three-element Windkessel model as the transpulmonary thermodilution model but rather uses a more sophisticated model that accounts for the finiteness of the pulse wave velocity and its reflection phenomena [55]. By providing a reference value for the cardiac output, systemic resistances are calculated during calibration. These resistances correspond to the ratio between the cardiac output and the mean arterial pressure, and the stroke volume is also reliably estimated during this calibration. The accuracy of the continuous cardiac output measurement is dependent on the iterative recalibrations. Indeed, it permits an understanding of the variations of the vasomotor tone, which are either spontaneous or induced by vasoactive drugs. Several studies

have compared thermodilution and CO measurements using the pulse curves analyses. A good correlation has been observed between all techniques, especially when patients had a stable vasomotor tone and received no or low doses of vasoactive drugs [24, 55, 56]. However, patients with arrhythmias (VES) or those whose blood pressure curves were not perfectly regular (artifacts) were excluded [57].

## 2.4 Pulse Contour Analysis Without Calibration

As observed with others technique, the calculation of cardiac output using the concept of the FloTrac/Vigileo™ system (Edwards Life Sciences, Irvine, California) is achieved by multiplying the stroke volume by the heart rate. The algorithm replaces the heart rate with the pulse. The stroke volume is calculated from the blood pressure using a system that analyzes the shape of the blood pressure wave via a proprietary algorithm. This statistical algorithm analyzes the pressure wave more than 100 times per second during a period of 20 s to capture more than 2,000 data points for analysis. These data points and the patient demographic informations are used for calculation of the standard deviation of the blood pressure (σAP). The σAP is proportional to the pulse pressure (PP). First, the σAP is multiplied by a known conversion factor (Chi ($\chi$)) that corresponds to the vascular tone. The σAP is initially expressed in mmHg but is converted to ml/beat. Thus, through the σAP and the vascular tone ($\chi$), the SV is calculated, beat by beat [58]:

$$CCO = HR \times (\sigma AP \times \chi)$$

$$\chi = M \left( HR, \sigma AP, C(P), BSA, MAP, \mu 3ap, \mu 4ap \ldots \right)$$

where CCO corresponds to the continuous cardiac output, HR corresponds to the heart rate, σAP corresponds to the standard deviation of the arterial pulse pressure expressed in mmHg, which is proportional to the pressure difference, $\chi$ corresponds to the multifactorial parameter proportional to the effects of vascular tone on the differential pressure, $M$ is the multifactorial polynomial equation, BSA corresponds to the body surface area calculated by the Dubois equation, MAP is the mean arterial pressure calculated by summing the pressure values sampled in 20 s and dividing by the number of pressure points, and $\mu$ corresponds to the statistical moments determined by skewness and kurtosis, as well as by various derived terms.

The algorithm applies the principle that the aortic pressure differential is proportional to the SV and inversely related to aortic compliance. Initially, the algorithm evaluates the pulse pressure using the standard deviation of the blood pressure (σAP) around the MAP value, measured in mmHg, making it independent of the effects of the vascular tone. The standard deviation of the differential pressure is proportional to the stroke volume and is recalculated 100 times per second for 20 s from the blood pressure wave, creating 2,000 data points from which the σAP is calculated. The standard deviation of the blood pressure, initially in mmHg

(converted to ml/beat), is calculated by multiplying by the conversion factor Chi ($\chi$). Chi is a multifactorial polynomial equation that estimates the effect of a variable vascular tone on the pulse pressure. Chi is calculated by analyzing the pulse rate, the mean arterial pressure, the standard deviation of the mean arterial pressure, the compliance of large vessels (estimated by patient demographics), asymmetry coefficients, and the flattening of the arterial wave. Chi is adjusted and inserted into the algorithm approximately every 60 s.

This system suffers from several limitations. A poor arterial pressure signal renders the analysis nonoperative. Although automatic corrections are made by the monitor, arrhythmias may occlude the regular ventricular premature beats and render the system nonoperative. Finally, in numerous validation studies, when the vascular tone was assigned a severe condition such as vasoplegia (septic shock) or when vasoactive drugs were infused, the cardiac output values provided by this technique lacked precision [59]. Therefore, the use of this technology in ICU patients in shock is cautioned and cannot be considered as a reference method [60].

Non-calibrated pulse contour analysis systems show acceptable precision in hemodynamically stable situations. In this review, 43 studies provided suitable data for aggregated and weighted analysis. The mean bias was significant (1.25 l/min) and indicated a percentage error of up to 40 %. In hemodynamically unstable patients, a higher error percentage was noted (up to 45 %) and a bias of more than 1.64 l/min was observed. Therefore, during an episode of hemodynamic instability, a CO measurement based on non-calibrated continuous pulse contour analyses show only limited correlation with a thermodilution intermittent bolus [61]. The calibrated systems appear to provide more accurate measurements than self-calibrated or non-calibrated systems [62]. For the reliable use of these semi-invasive systems, especially for critical treatment decisions during shock, it is necessary to define a hemodynamic optimization [63]. Others methods and techniques like the PRAM.

## 2.5  Other Techniques Using an Indicator Dilution

### 2.5.1  Indocyanine Green or Tricarbocyanine

For these techniques, the dye concentration is analyzed in the peripheral blood by a densitometer (specific wavelength) after an injection of approximately 5 mg in the pulmonary artery. This produces a curve representing the concentration of the dye as a function of time because the blood collection speed is constant (approximately 40 ml/min). CO is then calculated using the Stewart-Hamilton formula.

The area under the first movement time is calculated by extrapolating the descending portion. This makes it possible to account for the cardiac output in the calculation of the recirculation curve. However, similarly to all dilution techniques, in the case of low cardiac output or in the presence of severe valvular regurgitation or shunt, the measurement precision of the CO decreases. In addition, the relatively slow disappearance of the tracer makes it difficult to use in case of liver failure. Finally, allergic accidents also occur.

### 2.5.2  Lithium

For this technique, the plasma lithium concentration is measured after the injection of 0.6 mmol of lithium chloride through a venous catheter with an analytical system using a specific electrode. The present method is a transpulmonary thermodilution technique where the independent dilution technique is lithium chloride dilution using the Stewart-Hamilton principle. Lithium chloride dilution uses a peripheral vein and a peripheral arterial line. The injection of lithium chloride can be subject to errors in the presence of certain muscle relaxants [64]. Moreover, the potential for toxicity may compromise its use in the ICU.

## 2.6  Fick Methods

### 2.6.1  Conventional Method

The principle of this technique is based on the assumption that $O_2$ consumption ($VO_2$) is equal to the amount of $O_2$ added to the blood flowing through the lungs, using the following formula:

$$VO_2 = Q \times \left( CaO_2 - CvO_2 \right)$$

where $CaO_2$ and $CvO_2$ correspond to the $O_2$ contents of arterial and mixed venous blood, $Q$ corresponds to CO, and $VO_2$ corresponds to $O_2$ consumption.

The classical equation is as follows:

$$\dot{Q} = \frac{\dot{V}O_2}{CaO_2 - CvO_2}$$

$VO_2$ is then measured by the "gas exchange open circuit" method according to different techniques, as follows:

(a) A technique using a box or an indirect calorimetry chamber that does not require the separation of inspired and expired gas. This technique cannot be used in intubated and ventilated patients.

(b) A technique used on a patient on mechanical ventilation causes the separation of inspired and expired gases and the collection of exhaled gases inside a Douglas bag or a mixing chamber such that

$$\dot{V}O_2 = \dot{V}I \times FIO_2 - \dot{V}E \times FEO_2$$

where $\dot{V}I$ and $\dot{V}E$ are the inspired and expired flows, respectively, and FE and FI are the fractional concentrations of gas in inhaled or exhaled air, respectively. If limited to one of these two flows (usually $\dot{V}E$), then the hypothesis formulated by Haldane should be used, wherein the flow of nitrogen that enters the body ($\dot{V}I \times FIN_2$) is equal to the flow of nitrogen out of the body ($\dot{V}E \times FEN_2$):

$$FIN_2 + FIO_2 = 1$$

And

$$FEN_2 + FEO_2 + FECO_2 = 1$$

The following calculation formula is then obtained:

$$\dot{V}O_2 = \dot{V}E\left[\left(1 - FECO_2\right)FIO_2 - FEO_2\right)/\left(1 - FIO_2\right)\right]$$

There are several compact systems (e.g., SensorMedics, Engström Metabolic Computer, and Datex Deltatrac). The Deltatrac system measures the gas flow using a dilution technique.

Finally, measurement of the $O_2$ content requires the presence of a pulmonary artery catheter and a peripheral arterial catheter. Blood collection conditions require an adequate methodology. Particular attention should be paid to the sampling speed to ensure that the pulmonary artery catheter is in an unlocked position and to immediately analyze blood samples.

The Fick method is traditionally regarded as a reference method for CO measurement [65] in patients with hemodynamic and respiratory stability [66]. However, some authors have shown that this method is reliable and reproducible in hemodynamically unstable patients and during ventilatory weaning after cardiac surgery [67]. Conventional sources of error in the $VO_2$ measurement by this gas exchange method exist when FIO > 0.6 [68], due to the effects of the pressure and humidity of the gas analyzed [69] using halogenated gas and $N_2O$ and in the presence of chest drains. However, a precision of 4 % in the $VO_2$ measurement by this method is considered good [70]. The precision of CO measurement using the Fick principle is less than 5 % in normal subjects who are spontaneously breathing ambient air. This precision is satisfactory, with an error of approximately 3 % in the calculation of the arteriovenous $O_2$ difference

[65]. Several studies have compared the CO measured by thermodilution and calculated it according to the Fick principle [71–73]. The accuracy varies from one study to another (from 0.5 to 1.87 l/min). However, the average difference between the two methods is low. The lowest precision and bias values are obtained when the patient is placed in a controlled intermittent ventilation environment [67]. Other authors have shown that the reproducibility and accuracy of CO measurements by the Fick method are better than by thermodilution [84, 88]. In one study [84], the bias and precision were $0.1 \pm 1.1$ and $0.01 \pm 0.7$ l/min for CO values obtained by thermodilution and the Fick method, respectively. In cardiac surgery patients, the Deltatrac metabolism™ monitor using the Fick method to calculate the CO did not have sufficient accuracy to make this method a reference method [70]. This lack of precision may result from error accumulation in the $VO_2$ measurement and calculation of the arteriovenous difference in $O_2$.

### 2.6.2 CO$_2$ Consumption

#### 2.6.2.1 Classical Fick Method [66]
The principle of this method is to replace the $VO_2$ by $CO_2$ production ($VCO_2$) in the Fick equation:

$$\dot{Q} = \frac{\dot{V}CO_2}{1.34 \times Hb \times R \times \left(SaO_2 - SvO_2\right)}$$

where $R$ is the respiratory quotient and is considered constant with a value of 0.8 (or a value averaged over the measures).

However, these methods are highly dependent on the stability of the respiratory quotient and the measurement error of $SaO_2$ and $SvO_2$, whereas the partial pressure values are negligible. Therefore, variations in the CO obtained by this method are more correlated to those obtained by thermodilution or those measured by the classical Fick method [66].

## 2.6.3  CO₂ Rebreathing

The Fick principle can be applied to any gas diffusing through the lungs, including carbon dioxide. The NICO monitor (Novametrix Medical Systems, Inc., Wallingford, CT, USA) is based on application of the Fick principle to carbon dioxide to noninvasively estimate CO using intermittent partial rebreathing through a specific disposable rebreathing loop. The monitor comprises a carbon dioxide sensor (infrared light absorption), a disposable airflow sensor (differential pressure pneumotachometer), and a pulse oximeter. $VCO_2$ is calculated from minute ventilation and the carbon dioxide content, whereas the arterial carbon dioxide content ($CaCO_2$) is estimated from end-tidal carbon dioxide ($etCO_2$), with adjustments for the slope of the carbon dioxide dissociation curve and the degree of dead space ventilation. The partial rebreathing reduces carbon dioxide elimination and increases $etCO_2$ (equilibrium). Measurements under normal and rebreathing conditions allow one to omit the venous carbon dioxide content ($CvCO_2$) measurement in the Fick equation; therefore, the need for a central venous access is eliminated. The principle used by the NICO monitor is as follows.

Fick equation applied to carbon dioxide:

$$CO = \frac{VCO_2}{CvCO_2 - CaCO_2}$$

Assuming that cardiac output remains unchanged under normal ($N$) and rebreathing ($R$) conditions,

$$CO = \frac{VCO_{2N}}{CvCO_{2N} - CaCO_{2N}} = \frac{VCO_{2R}}{CvCO_{2R} - CaCO_{2R}}$$

By subtracting the normal and rebreathing ratios, the following differential Fick equation is obtained:

$$CO = \frac{VCO_{2N} - VCO_{2R}}{\left( CvCO_{2N} - CaCO_{2N} \right) - \left( CvCO_{2R} - CaCO_{2R} \right)}$$

Because carbon dioxide diffuses quickly in the blood (22 times faster than oxygen), one can assume that the $CvCO_2$ does not differ between normal and rebreathing conditions; therefore, the venous contents disappear from the equation:

$$CO = \frac{\Delta VCO_2}{\Delta CaCO_2}$$

The delta in $CaCO_2$ can be approximated by the delta in $etCO_2$ multiplied by the slope ($S$) of the carbon dioxide dissociation curve. This curve represents the relationship between carbon dioxide volumes (used to calculate the carbon dioxide content) and the partial pressure of carbon dioxide. This relationship can be considered linear between 15 and 70 mmHg partial pressure values of carbon dioxide [74].

$$CO = \frac{\Delta VCO_2}{S \times \Delta etCO_2}$$

Because changes in $VCO_2$ and $etCO_2$ only reflect the blood flow that participates in gas exchange, an intrapulmonary shunt can affect estimation of the cardiac output using the NICO device. To account for this effect, the monitor estimates the shunting fraction using a measured peripheral oxygen saturation of hemoglobin combined with the $FiO_2$ and the arterial oxygen tension measured in arterial blood gases according to Nunn's iso-shunt tables [75].

Increased intrapulmonary shunt and poor hemodynamic stability (which are not uncommon in critically ill patients) are likely to alter the precision of cardiac output estimation by the NICO monitor. The first published clinical and experimental validation studies [76–78] reported a relatively loose agreement (bias ± 1.8 l/min) between the CO measured using thermodilution and that measured using the NICO device (this is similar to standard observations whenever a technique is compared with thermodilution). Those investigators therefore concluded that the technique is not yet ready to replace the thermodilution technique. However, comparable limits of agreements have been observed in many studies that compared cardiac output measurement techniques with thermodilution, including "bolus" versus "continuous" thermodilution [79–81].

Bland and Altman [82] asserted that a tight agreement is impossible to obtain when the method used for reference itself is not very precise. Such limits of agreement do not preclude the potential usefulness of cardiac output measurement using the NICO monitor; however, the abovementioned limitations must be considered, and the technique should be used only in the most appropriate patients.

Notably, the patient must be under fully controlled mechanical ventilation if the NICO monitor is to be used. In addition, arterial blood samples are required to enter arterial oxygen tension values for shunt estimation, which somewhat tempers the noninvasive nature of this technique.

### 2.6.4 Soluble Inert Gas

CO measurement method using the acetylene rebreathing method based on the Fick principle remains the dominant method [83]. There are many sources of error (e.g., intrapulmonary shunt, abnormal ventilation-perfusion ratio, and recirculation) that do not permit its use in clinical practice in the ICU.

## 2.7 Doppler Methods

### 2.7.1 Methods

The flow rate (velocimetry) of blood in the vessels is measured by the Doppler effect. This technique uses ultrasound waves emitted by a probe, which then spread into the soft tissue toward the tissue to be studied. These waves encounter a moving blood column in which the red blood cells produce a distribution of ultrasonic acoustic energy. The portion of the waves scattered back is then detected by the probe. The frequency of the ultrasonic signal received by the probe differs from the frequency ($F$) of the transmitted signal by a delta-$E$ value ($dF$) due to the Doppler effect. The mathematical relationship among the ultrasonic propagation velocity in the soft tissue ($C = 1,540$ m/s), the velocity ($V$)

of the erythrocytes, $dF$, $F$, and the angle of incidence ($A$) is expressed as follows:

$$dF = \frac{2F \times V \times \mathrm{Cos}A}{C}$$

The Doppler effect is first observed between the probe (fixed component) and the red blood cells (mobile receivers) and then between the red blood cells (mobile transmitters) and the probe (fixed receiver) during backscattering, explaining the presence of a coefficient of 2. In view of both the usual transmission frequency between 2 and 10 MHz and flow velocities in the main vessels of the body, the $dF$ frequency is generally between 100 and 20,000 Hz, corresponding to the auditory field. It is sufficient to amplify the Doppler frequency for locating and identifying vessels and for analyzing the circulatory conditions. Finally, to remove the low frequencies from the pulsatile motion of the vessel walls or the movement of the probe, the Doppler signal can be filtered by the device.

### 2.7.2 Continuous or Pulsed Doppler

#### 2.7.2.1 Continuous Doppler

The probe comprises two permanent transducers: a transmitter and a receiver. This is a simple and inexpensive technology with a good signal to noise ratio. However, it has no spatial resolution; all flows encountered by the ultrasound beam along its path are considered. For example, the continuous Doppler wave records signals from a target artery and its adjacent vein. The maximal velocity is then measured.

#### 2.7.2.2 Pulsed Doppler

The pulsed Doppler signal permits spatial resolution. It is possible to select vessels according to their topography and regions of space in which the Doppler signal is obtained. The probe is provided with a single transducer, which alternately acts as a transmitter and a receiver. The transmission is made by short pulses on the order of microseconds. The sensor operates in the receiving mode between two successive pulses (reception window) and saves the Doppler signals from impulse. Adjustment of the reception window is the boundary of a "measurement volume," which corresponds to the region of the space in which the Doppler signals are collected. Adjustment of the window is facilitated by the use of ultrasound to measure the distance between the source and the blood vessel studied. "Windowing" pulsed Doppler systems are able to perform flow velocity measurements in a large number of points along the Doppler signal line. It is possible to raise the profiles of flow velocities. These devices are the basis of dynamic mapping systems or "color Doppler." Despite the major advantage of the spatial resolution, pulsed Doppler suffers from several drawbacks: it is a complex and expensive technique that requires an experienced operator. The signal to noise ratio is less favorable than that of continuous Doppler, but its instantaneous acoustic powers are higher.

### 2.7.2.3 Blood Flow Measurement

Doppler assesses the blood velocity during systole through a surface area ($s(t)$) over time. The instantaneous flow rate is then calculated by the following formula:

$$CO = v(t) \times s(t)$$

where $v(t)$ corresponds to the average speed of the blood column at time $t$, expressed in m/s, and $s(t)$ corresponds to the section of the vessel as a function of time, expressed in m$^2$. When applied to the heart chamber or to the aorta, the above formula is used to measure the volume through the studied section such that

$$SV = \sum_{o}^{t} v(t) \times s(t) \times dt$$

where SV corresponds to the stroke volume ($I$) and t is the study time in seconds. If the section ($S$) is considered constant, the equation becomes:

$$SV = S \sum_{o}^{t} v(t) \times dt$$

The integral of $v(t)$ on a systole represents the area under the curve of the blood flow velocities (stroke distance in cms). This surface is measured

by planimetry or if circular using ring surface equation. CO is then

$$CO = HR \times S \sum_{o}^{t} v(t) \times dt$$

To measure CO, several conditions must be met:

(a) Ideally, the blood flow must be a laminar flow through the measurement section.
(b) The blood velocity should be uniform throughout this section.
(c) The angle of incidence of Doppler firing must be known.
(d) Measuring the average section of the circle is necessary because vessels and cardiac chambers undergo deformations during systole and diastole.

## 2.8 Doppler Methods for the Measurement of Cardiac Output

### 2.8.1 Echocardiography

Transthoracic Doppler echocardiography is used to measure blood flow velocities on the mitral and aortic valves. Various studies have shown a good performance of this method [84, 85]. However, the ultrasonic beam can be reduced in cases of prosthetic valves, pulmonary emphysema, pneumothorax, and COPD, as well as during mechanical ventilation, particularly after cardiothoracic surgery. Bidimensional echocardiography performed by the transesophageal route and coupled with the Doppler method also allows for CO measurement. CO estimation from the SV can be conducted through the mitral, aortic, or pulmonary valves [86]. The use of a biplane probe allows for correct placement of the Doppler beam at the outflow tract of the left ventricle or mitral valve. This method then provides a good estimate of the CO compared with the thermodilution method [87]. However, echocardiography requires an expensive apparatus. In addition, its widespread use in intensive care for the most severe patients is often limited to short periods, particularly during the perioperative period. Finally, echocardiography requires a significant learning curve and qualified operators. Indeed, it is estimated that more than 100 transesophageal echocardiography cases are necessary to qualify an operator [88].

## 2.8.2 Suprasternal Doppler

The use of suprasternal Doppler allows for the estimation of blood velocity in either the ascending aorta or the aortic arch [89]. This technique uses continuous or pulsed emission transducers. The ultrasound probes are placed at the suprasternal notch. By searching for the maximum flow, it is possible to obtain the velocity curves to estimate the SV. The ultrasound beam should be oriented to find the flow of the ascending aorta or the flow of the aortic arch. The diameter of the aorta is obtained either by ultrasound or by using a preestablished nomogram [90]. To obtain a reliable measurement of the CO by this technique, it is necessary to average at least five areas because of breathing-related changes [91]. Despite the ease and speed of access and the noninvasive nature of the technique, this method can be used in only 5 % of patients. Anatomical conditions (e.g., short neck) and diseases (e.g., emphysema, mediastinal air after cardiac surgery, aortic valve pathology) make the application of this technology unrealistic

in clinical practice [92]. In addition, this technique provides information on left ventricular function by measuring the maximum velocity, as well as the acceleration if there is no infringement of the aortic valve [93] (Fig. 2.13). Also, estimation of the preload and the afterload may be possible by analyzing the shape of the curve obtained. However, as is the case with echocardiography, continuous measurement of the different parameters is not feasible.

## 2.8.3 Transtracheal Doppler

This method uses a transducer that is placed at the end of an endotracheal tube to measure the diameter and blood velocity in the ascending aorta [94, 95]. In addition to the high price of the technique (the endotracheal tube is disposable, and its price is much higher than a standard intubation probe), the reliability of this technique is debatable in the ICU [95]. Although the accuracy of the CO measurement is not optimal, it improves with practice by the operator [96].

## 2.8.4 Esophageal Doppler

This method measures the blood velocity in the descending thoracic aorta using a probe placed in

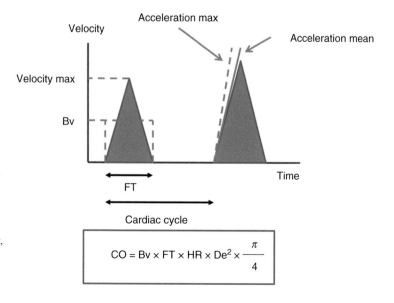

**Fig. 2.13** Blood velocity curves measured by Doppler and calculation of the cardiac output by the suprasternal Doppler method. *Flow time (FT)* ejection duration, *Bv* mean blood velocity, *De* (nomogram) effective aortic diameter, *HR* heart rate, and *CO* cardiac output

$$CO = Bv \times FT \times HR \times De^2 \times \frac{\pi}{4}$$

the esophagus (between the fifth and sixth inter-costal space) [92]. After determining the diameter of the aorta, the blood velocity is measured by the Doppler method (continuous or pulsed). At the level of the esophagus, the sources of errors related to the nature of the vessel studied are minimal. The aortic signal is easily differentiated from the inferior vena cava signal [92].

To reliably measure aortic flow, a number of requirements must be met [92].

The angle of incidence between the direction of flow and the ultrasound beam must be correct and must remain constant throughout the monitoring period (especially during mechanical ventilation). The formula for calculating the blood flow velocity uses the cosine of the angle of incidence of the ultrasonic beam on the studied vessel. At greater incidence angles, the risk of error in the CO measurement increases. The esophageal Doppler devices must use angles of incidence between 45° and 60° because the esophagus and aorta are parallel. An error of 5° causes a CO measurement error of more than 10 %.

Despite variations in blood pressure and CO, the aortic surface must remain constant during systole. Measurement of the aortic diameter by time-motion (TM) echocardiography may be underestimated if the ultrasonic beam does not cut the aorta in the center. Conversely, the aortic diameter is overestimated when the beam is not perfectly perpendicular to the axis of the aorta. A two-dimensional ultrasound faces the same risks. The area of a circle is the transformed surface of an ellipse. In other devices, algorithms are used to estimate the diameter of the aorta according to the age, weight, size, and sex of the patient and are sometimes adjusted to the MAP [97]. However, there is a risk of error in the determination of these parameters, which then automatically introduces error in the estimation of the aortic flow and CO.

The flow distribution between the descending thoracic aorta (approximately 70 %) and the coronary and carotid arteries (approximately 30 %) remains constant (apportionment factor $K$) regardless of the clinical situation. Therefore, the flow distribution is considered constant during measurements; this factor may be determined at

the beginning of the measurement of aortic flow. However, differences between thermodilution measurement of the aortic blood flow and the CO in aortic clamping can be explained by changes in this factor [98].

The method of measuring CO by transesophageal Doppler remains imprecise compared with the thermodilution method despite recent technological improvements [109]. However, significant variations in the CO measured by the two methods are often correlated [92, 97, 99].

Doppler methods cause many errors in CO measurement. They do not allow accurate assessment of the CO in intensive care. The addition of Doppler ultrasound imaging to these methods is sometimes used to guide the diagnosis in cases of circulatory failure. However, although the learning time of transesophageal ultrasound is small compared with that required to learn the practice of conventional ultrasound, all of these techniques remain very dependent on their operators. Finally, frequent repositioning of the esophageal probe is necessary, and the patient must stay perfectly immobile. Therefore, these factors make the transesophageal Doppler method a bit simple, unreliable, and poorly reproducible for continuous CO measurements in the ICU.

## 2.9    Thoracic Bioimpedance

CO measurement by the thoracic bioimpedance method is based on mathematical analysis of the variations of the consecutive transthoracic resistance to changes in intrathoracic blood volume by applying an alternating current of low amplitude and high frequency. Pulsatile increases in the intrathoracic blood volume at each systole decrease the pulsatility in chest impedance due to the good conductivity of blood. Several mathematical formulas are used to calculate SV:

$$SV = r^3 \left( L / Z_0 \right)^2 3t3dZ / dt_{max} \quad [100]$$

$$SV = \frac{L^3 \times LVET \times dZ / dt_{max}}{4.25 \times Z_0} \quad [101]$$

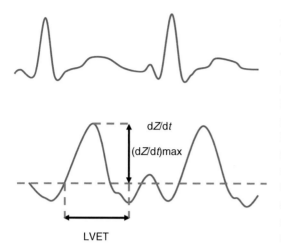

**Fig. 2.14** Thoracic impedance curve [101]. *LVET* corresponds to the left ventricular ejection time, and *dZ/dt* is the maximum variation of thoracic impedance

where $r$ corresponds to the resistivity of the blood, expressed in Ohm/cm; $L$ corresponds to the rib length, i.e., the distance between the internal electrodes or nomogram; $Z_0$ corresponds to the initial impedance; $t$ or LVET corresponds to the left ventricular ejection time; and $dZ/dt_{max}$ corresponds to the maximum variation in impedance during systole, which is a reflection of the rapid ejection phase in early systole (Fig. 2.14).

The thorax is considered a cylinder in the first pattern [100], whereas it is considered to be a truncated cone in the second model [101]. In both models, the velocity during the ejection is considered constant.

Using remotely placed (e.g., in the neck or chest) electrodes (patch or tape), variable alternating currents (100 or 200 Hz, 2.5–4 mA) are applied. The impedance variations are then detected via surface electrodes, which are placed at a distance from the first electrodes so that they produce a current through the thorax in a direction that is parallel to the spine. A curve is then recorded. The ECG signal enables the computer to determine the beginning of systole in patients who have an average impedance signal.

Most studies comparing this technique with thermodilution show mean differences and accreditation boundaries that are clinically acceptable [95, 102, 103]. Variations in the CO

measured by impedance have been both unpredictable or properly predicted [95, 103, 104] by thermodilution. Errors in CO measurement are dependent on the method, the patient, and the environment [105]. An insufficient distance between the receiving electrodes overestimates the CO. The opposite occurs when the distance between the electrodes is too large. An error in the size and weight of the patient is likely to cause an error of 10–30 % in CO estimation. Obesity, mechanical ventilation, pulmonary edema or pleural effusions, and rhythm disorders can cause a decrease in the accuracy of the measurement [103]. CO measurement by bioimpedance is overestimated in the case of low CO and is underestimated in the case of high CO when this method is compared with thermodilution.

In conclusion, although bioimpedance is interesting under stable physiological conditions, it seems ill-suited for hemodynamically unstable ICU patients [106, 107].

## 2.10   Other Methods of Measuring Cardiac Output

There are numerous other methods for measuring cardiac output; however, some are not frequently used in clinical practice, and others are under development.

### 2.10.1 Method According to a Flow Model

Cardiac output is derived using the so-called Modelflow method (simulation of a three-element Windkessel model). Regarding validation of the device, only limited published data are available [108].

To sum up the present paragraph related to cardiac output measurements, we may highlight that, in some studies, the pulmonary artery catheter use might be not only ineffective but sometimes also potentially deleterious [109]. This notion, together with the availability of new, less invasive CO monitoring devices, has markedly decreased PAC use [110]. Today, several devices are available to measure or estimate CO using different methods. However, notably, each device has inherent limitations, and no CO monitoring device can change patient outcomes unless its use is coupled with an intervention that alone has been associated with improved patient outcomes. Therefore, the concept of hemodynamic optimization is increasingly recognized as a cornerstone in the management of critically ill patients. Hemodynamic optimization is associated with improved outcome in the perioperative period [111] and in the ICU [112] setting. When choosing a CO monitoring device for bedside use, various factors must be considered. Institutional factors may largely limit the choice of the available devices, and important device-related factors may restrict the area of application. Furthermore, patient-specific conditions may dictate the use of an invasive or a minimally or noninvasive device.

Considering the technical features and typical limitations of the different CO monitoring techniques, it is obvious that no single device can

comply with all clinical requirements. Therefore, different devices may be used in an integrative concept along a typical clinical patient pathway (Fig. 2.15) based on the invasiveness of the devices and the available additional hemodynamic variables. Bioreactance may be used on the ward or in the emergency department to initially assess CO to confirm a preliminary diagnosis. Its use may be expanded in the perioperative and ICU settings. Partial $CO_2$ rebreathing requires an intubated and mechanically ventilated patient for cardiac output estimation but may be of interest inside an ambulance of emergency medical service. Uncalibrated pulse pressure analysis devices may be the primary choice in a perioperative setting as they provide trend values of functional hemodynamic variables and thus allow comprehensive hemodynamic management. In contrast, calibrated systems may be required when postoperative complications or hemodynamic instability occur and when increased device accuracy or volumetric variables are required for improved patient management like

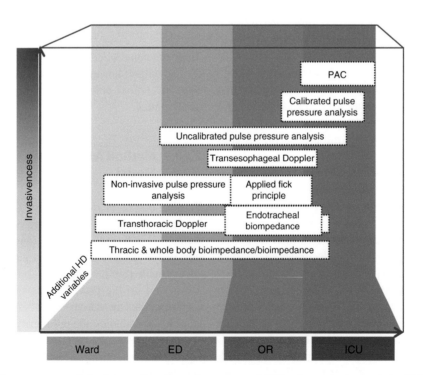

**Fig. 2.15** Integrative concept for the use of cardiac output monitoring devices. *ED* emergency department, *HD* hemodynamic, *ICU* intensive care unit, *OR* operating room, and *PAC* pulmonary artery catheter (Adapted from Alhashemi et al. [113])

in the ICU. In the presence of factors that affect the accuracy of all minimally invasive cardiac output monitoring devices or when pulmonary artery pressure monitoring or right heart failure treatment is required, PAC insertion may be required for patient-specific therapy [113].

To conclude this section on CO measurement, various devices that allow for continuous cardiac output measurement in critically ill patients are commercially available today. Their presence does not completely preclude but does increasingly limit PAC use. A variety of factors (institutional, device related, and patient specific) influence the selection of a cardiac output monitoring device, and clinicians must understand the underlying principles and inherent limitations of these devices. A selection of these techniques may be used in an integrative approach along a critically ill patient pathway.

## References

1. Harvey S, Young D, Brampton W, Cooper AB, Doig G, Sibbald W et al (2006) Pulmonary artery catheters for adult patients in intensive care. Cochrane Database Syst Rev (3):CD003408. [Meta-Analysis Review]
2. Shah MR, Hasselblad V, Stevenson LW, Binanay C, O'Connor CM, Sopko G et al (2005) Impact of the pulmonary artery catheter in critically ill patients: meta-analysis of randomized clinical trials. JAMA J Am Med Assoc 294(13):1664–1670
3. Jansen JR (1995) The thermodilution method for the clinical assessment of cardiac output. Intensive Care Med 21(8):691–697
4. Rubini A, Del Monte D, Catena V, Attar I, Cesaro M, Soranzo D et al (1995) Cardiac output measurement by the thermodilution method: an in vitro test of accuracy of three commercially available automatic cardiac output computers. Intensive Care Med 21(2):154–158
5. Levett JM, Replogle RL (1979) Thermodilution cardiac output: a critical analysis and review of the literature. J Surg Res 27(6):392–404
6. Stetz CW, Miller RG, Kelly GE, Raffin TA (1982) Reliability of the thermodilution method in the determination of cardiac output in clinical practice. Am Rev Respir Dis 126(6):1001–1004
7. Renner LE, Morton MJ, Sakuma GY (1993) Indicator amount, temperature, and intrinsic cardiac output affect thermodilution cardiac output accuracy and reproducibility. Crit Care Med 21(4):586–597, Research Support, Non-U.S. Gov't Research Support, U.S. Gov't, P.H.S
8. Latson TW, Whitten CW, O'Flaherty D (1993) Ventilation, thermal noise, and errors in cardiac output measurements after cardiopulmonary bypass. Anesthesiology 79(6):1233–1243
9. Synder JV, Powner DJ (1982) Effects of mechanical ventilation on the measurement of cardiac output by thermodilution. Crit Care Med 10(10):677–682
10. Jansen JR, Schreuder JJ, Settels JJ, Kloek JJ, Versprille A (1990) An adequate strategy for the thermodilution technique in patients during mechanical ventilation. Intensive Care Med 16(7):422–425
11. Assmann R, Heidelmeyer CF, Trampisch HJ, Mottaghy K, Versprille A, Sandmann W et al (1991) Right ventricular function assessed by thermodilution technique during apnea and mechanical ventilation. Crit Care Med 19(6):810–817
12. Sasse SA, Chen PA, Berry RB, Sassoon CS, Mahutte CK (1994) Variability of cardiac output over time in medical intensive care unit patients. Crit Care Med 22(2):225–232
13. Yelderman M (1990) Continuous measurement of cardiac output with the use of stochastic system identification techniques. J Clin Monit 6(4):322–332
14. Bizouarn P, Blanloeil Y, Pinaud M (1994) Comparison of cardiac output measured continuously by thermodilution and calculated according to Fick's principle. Ann Fr Anesth Reanim 13(5):685–689 [Comparative Study]
15. Boldt J, Menges T, Wollbruck M, Hammermann H, Hempelmann G (1994) Is continuous cardiac output measurement using thermodilution reliable in the critically ill patient? Crit Care Med 22(12):1913–1918
16. Haller M, Zollner C, Briegel J, Forst H (1995) Evaluation of a new continuous thermodilution cardiac output monitor in critically ill patients: a prospective criterion standard study. Crit Care Med 23(5):860–866
17. Jakobsen CJ, Melsen NC, Andresen EB (1995) Continuous cardiac output measurements in the perioperative period. Acta Anaesthesiol Scand 39(4):485–488
18. Lefrant JY, Bruelle P, Ripart J, Ibanez F, Aya G, Peray P et al (1995) Cardiac output measurement in critically ill patients: comparison of continuous and conventional thermodilution techniques. Can J Anaesth J Can Anaesth 42(11):972–976
19. Mihaljevic T, von Segesser LK, Tonz M, Leskosek B, Seifert B, Jenni R et al (1995) Continuous versus bolus thermodilution cardiac output measurements – a comparative study. Crit Care Med 23(5):944–949 [Comparative Study]
20. Bendjelid K, Schutz N, Suter PM, Romand JA (2006) Continuous cardiac output monitoring after cardiopulmonary bypass: a comparison with bolus thermodilution measurement. Intensive Care Med 32(6):919–922 [Comparative Study]
21. Nelson LD (1996) The new pulmonary arterial catheters. Right ventricular ejection fraction and continuous cardiac output. Crit Care Clin 12(4):795–818

22. Haddad F, Couture P, Tousignant C, Denault AY
    (2009) The right ventricle in cardiac surgery, a peri-
    operative perspective: I. Anatomy, physiology, and
    assessment. Anesth Analg 108(2):407–421
23. Diebel LN, Wilson RF, Tagett MG, Kline RA (1992)
    End-diastolic volume. A better indicator of preload
    in the critically ill. Arch Surg 127(7):817–821;
    discussion 21–22
24. Goedje O, Hoeke K, Lichtwarck-Aschoff M,
    Faltchauser A, Lamm P, Reichart B (1999)
    Continuous cardiac output by femoral arterial ther-
    modilution calibrated pulse contour analysis: com-
    parison with pulmonary arterial thermodilution. Crit
    Care Med 27(11):2407–2412 [Comparative Study]
25. Goedje O, Peyerl M, Seebauer T, Dewald O,
    Reichart B (1998) Reproducibility of double indi-
    cator dilution measurements of intrathoracic blood
    volume compartments, extravascular lung water, and
    liver function. Chest 113(4):1070–1077
26. Goedje O, Seebauer T, Peyerl M, Pfeiffer UJ,
    Reichart B (2000) Hemodynamic monitoring by
    double-indicator dilution technique in patients after
    orthotopic heart transplantation. Chest 118(3):775–
    781 [Comparative Study]
27. Sakka SG, Reinhart K, Meier-Hellmann A (1999)
    Comparison of pulmonary artery and arterial ther-
    modilution cardiac output in critically ill patients.
    Intensive Care Med 25(8):843–846 [Clinical Trial
    Comparative Study]
28. Tibby SM, Hatherill M, Marsh MJ, Morrison G,
    Anderson D, Murdoch IA (1997) Clinical validation
    of cardiac output measurements using femoral artery
    thermodilution with direct Fick in ventilated chil-
    dren and infants. Intensive Care Med 23(9):987–991
    [Comparative Study]
29. Pauli C, Fakler U, Genz T, Hennig M, Lorenz HP,
    Hess J (2002) Cardiac output determination in chil-
    dren: equivalence of the transpulmonary thermodi-
    lution method to the direct Fick principle. Intensive
    Care Med 28(7):947–952
30. Meier P, Zierler KL (1954) On the theory of the
    indicator-dilution method for measurement of blood
    flow and volume. J Appl Physiol 6(12):731–744
31. Newman EV, Merrell M, Genecin A, Monge C,
    Milnor WR, Mc KW (1951) The dye dilution
    method for describing the central circulation. An
    analysis of factors shaping the time-concentration
    curves. Circulation 4(5):735–746
32. Sakka SG, Ruhl CC, Pfeiffer UJ, Beale R, McLuckie
    A, Reinhart K et al (2000) Assessment of cardiac
    preload and extravascular lung water by single
    transpulmonary thermodilution. Intensive Care Med
    26(2):180–187
33. Hofer CK, Furrer L, Matter-Ensner S, Maloigne
    M, Klaghofer R, Genoni M et al (2005) Volumetric
    preload measurement by thermodilution: a com-
    parison with transoesophageal echocardiography. Br
    J Anaesth 94(6):748–755
34. Michard F, Alaya S, Zarka V, Bahloul M, Richard C,
    Teboul JL (2003) Global end-diastolic volume as an
    indicator of cardiac preload in patients with septic
    shock. Chest 124(5):1900–1908
35. Michard F, Schachtrupp A, Toens C (2005) Factors
    influencing the estimation of extravascular lung
    water by transpulmonary thermodilution in critically
    ill patients. Crit Care Med 33(6):1243–1247
36. Nirmalan M, Willard TM, Edwards DJ, Little RA,
    Dark PM (2005) Estimation of errors in determining
    intrathoracic blood volume using the single transpul-
    monary thermal dilution technique in hypovolemic
    shock. Anesthesiology 103(4):805–812
37. Reuter DA, Felbinger TW, Moerstedt K, Weis F,
    Schmidt C, Kilger E et al (2002) Intrathoracic blood
    volume index measured by thermodilution for pre-
    load monitoring after cardiac surgery. J Cardiothorac
    Vasc Anesth 16(2):191–195
38. Wiesenack C, Prasser C, Keyl C, Rodig G (2001)
    Assessment of intrathoracic blood volume as an indi-
    cator of cardiac preload: single transpulmonary ther-
    modilution technique versus assessment of pressure
    preload parameters derived from a pulmonary artery
    catheter. J Cardiothorac Vasc Anesth 15(5):584–588
39. Katzenelson R, Perel A, Berkenstadt H, Preisman S,
    Kogan S, Sternik L et al (2004) Accuracy of trans-
    pulmonary thermodilution versus gravimetric mea-
    surement of extravascular lung water. Crit Care Med
    32(7):1550–1554
40. Kirov MY, Kuzkov VV, Kuklin VN, Waerhaug
    K, Bjertnaes LJ (2004) Extravascular lung water
    assessed by transpulmonary single thermodilution
    and postmortem gravimetry in sheep. Crit Care
    8(6):R451–R458
41. Sakka SG, Klein M, Reinhart K, Meier-Hellmann A
    (2002) Prognostic value of extravascular lung water
    in critically ill patients. Chest 122(6):2080–2086
42. Ware LB, Matthay MA (2005) Clinical prac-
    tice. Acute pulmonary edema. N Engl J Med
    353(26):2788–2796
43. Richard C, Teboul JL (2005) Weaning failure
    from cardiovascular origin. Intensive Care Med
    31(12):1605–1607
44. Lemaire F, Teboul JL, Cinotti L, Giotto G, Abrouk F,
    Steg G et al (1988) Acute left ventricular dysfunction
    during unsuccessful weaning from mechanical ven-
    tilation. Anesthesiology 69(2):171–179
45. Kuzkov VV, Kirov MY, Sovershaev MA, Kuklin VN,
    Suborov EV, Waerhaug K et al (2006) Extravascular
    lung water determined with single transpulmo-
    nary thermodilution correlates with the severity of
    sepsis-induced acute lung injury. Crit Care Med
    34(6):1647–1653 [Comparative Study Research
    Support, Non-U.S. Gov't]
46. Monnet X, Anguel N, Osman D, Hamzaoui O,
    Richard C, Teboul JL (2007) Assessing pulmo-
    nary permeability by transpulmonary thermo-
    dilution allows differentiation of hydrostatic
    pulmonary edema from ALI/ARDS. Intensive Care
    Med 33(3):448–453
47. Giraud R, Siegenthaler N, Park C, Beutler S,
    Bendjelid K (2010) Transpulmonary thermodilution

curves for detection of shunt. Intensive Care Med 36(6):1083–1086

48. Martin GS, Eaton S, Mealer M, Moss M (2005) Extravascular lung water in patients with severe sepsis: a prospective cohort study. Crit Care 9(2):R74–R82 [Comparative Study Research Support, N.I.H., Extramural Research Support, Non-U.S. Gov't Research Support, U.S. Gov't, P.H.S.]

49. Robotham JL, Takata M, Berman M, Harasawa Y (1991) Ejection fraction revisited. Anesthesiology 74(1):172–183

50. Combes A, Berneau JB, Luyt CE, Trouillet JL (2004) Estimation of left ventricular systolic function by single transpulmonary thermodilution. Intensive Care Med 30(7):1377–1383

51. Jabot J, Monnet X, Bouchra L, Chemla D, Richard C, Teboul JL (2009) Cardiac function index provided by transpulmonary thermodilution behaves as an indicator of left ventricular systolic function. Crit Care Med 37(11):2913–2918

52. Berkenstadt H, Margalit N, Hadani M, Friedman Z, Segal E, Villa Y et al (2001) Stroke volume variation as a predictor of fluid responsiveness in patients undergoing brain surgery. Anesth Analg 92(4):984–989

53. Reuter DA, Felbinger TW, Schmidt C, Kilger E, Goedje O, Lamm P et al (2002) Stroke volume variations for assessment of cardiac responsiveness to volume loading in mechanically ventilated patients after cardiac surgery. Intensive Care Med 28(4):392–398

54. Linton R, Band D, O'Brien T, Jonas M, Leach R (1997) Lithium dilution cardiac output measurement: a comparison with thermodilution. Crit Care Med 25(11):1796–1800

55. Linton NW, Linton RA (2001) Estimation of changes in cardiac output from the arterial blood pressure waveform in the upper limb. Br J Anaesth 86(4):486–496

56. Zollner C, Haller M, Weis M, Morstedt K, Lamm P, Kilger E et al (2000) Beat-to-beat measurement of cardiac output by intravascular pulse contour analysis: a prospective criterion standard study in patients after cardiac surgery. J Cardiothorac Vasc Anesth 14(2):125–129

57. Berberian G, Quinn TA, Vigilance DW, Park DY, Cabreriza SE, Curtis LJ et al (2005) Validation study of PulseCO system for continuous cardiac output measurement. ASAIO J 51(1):37–40

58. Giraud R, Siegenthaler N, Bendjelid K (2011) Pulse pressure variation, stroke volume variation and dynamic arterial elastance. Crit Care 15(2):414

59. Desebbe O, Henaine R, Keller G, Koffel C, Garcia H, Rosamel P et al (2013) Ability of the third-generation FloTrac/Vigileo software to track changes in cardiac output in cardiac surgery patients: a polar plot approach. J Cardiothorac Vasc Anesth 27(6):1122–1127

60. Suehiro K, Tanaka K, Funao T, Matsuura T, Mori T, Nishikawa K (2013) Systemic vascular resistance has an impact on the reliability of the Vigileo-FloTrac system in measuring cardiac output and tracking cardiac output changes. Br J Anaesth 111(2):170–177 [Research Support, Non-U.S. Gov't]

61. Peyton PJ, Chong SW (2010) Minimally invasive measurement of cardiac output during surgery and critical care: a meta-analysis of accuracy and precision. Anesthesiology 113(5):1220–1235

62. Palmers PJ, Vidts W, Ameloot K, Cordemans C, Van Regenmortel N, De Laet I et al (2012) Assessment of three minimally invasive continuous cardiac output measurement methods in critically ill patients and a review of the literature. Anaesthesiol Intensiv Ther 44(4):188–199

63. Schloglhofer T, Gilly H, Schima H (2014) Semi-invasive measurement of cardiac output based on pulse contour: a review and analysis. Can J Anaesth J Can Anaesth 61(5):452–479

64. Linton RA, Band DM, Haire KM (1993) A new method of measuring cardiac output in man using lithium dilution. Br J Anaesth 71(2):262–266

65. Taylor SH (1966) Measurement of the cardiac output in man. Proc R Soc Med 59(Suppl):35–53

66. Mahutte CK, Jaffe MB, Chen PA, Sasse SA, Wong DH, Sassoon CS (1994) Oxygen Fick and modified carbon dioxide Fick cardiac outputs. Crit Care Med 22(1):86–95

67. Keinanen O, Takala J, Kari A (1992) Continuous measurement of cardiac output by the Fick principle: clinical validation in intensive care. Crit Care Med 20(3):360–365 [Comparative Study]

68. Ultman JS, Bursztein S (1981) Analysis of error in the determination of respiratory gas exchange at varying FIO2. J Appl Physiol Respir Environ Exerc Physiol 50(1):210–216

69. Takala J, Keinanen O, Vaisanen P, Kari A (1989) Measurement of gas exchange in intensive care: laboratory and clinical validation of a new device. Crit Care Med 17(10):1041–1047

70. Bizouarn P, Blanloeil Y, Pinaud M (1995) Comparison between oxygen consumption calculated by Fick's principle using a continuous thermodilution technique and measured by indirect calorimetry. Br J Anaesth 75(6):719–723 [Clinical Trial Comparative Study Controlled Clinical Trial Research Support, Non-U.S. Gov't]

71. Carpenter JP, Nair S, Staw I (1985) Cardiac output determination: thermodilution versus a new computerized Fick method. Crit Care Med 13(7):576–579

72. Mahutte CK, Jaffe MB, Sassoon CS, Wong DH (1991) Cardiac output from carbon dioxide production and arterial and venous oximetry. Crit Care Med 19(10):1270–1277

73. Quinn TJ, Weissman C, Kemper M (1991) Continual trending of Fick variables in the critically ill patient. Chest 99(3):703–707

74. McHardy GJ (1967) The relationship between the differences in pressure and content of carbon dioxide in arterial and venous blood. Clin Sci 32(2):299–309

75. Benatar SR, Hewlett AM, Nunn JF (1973) The use of iso-shunt lines for control of oxygen therapy. Br J Anaesth 45(7):711–718

76. van Heerden PV, Baker S, Lim SI, Weidman C, Bulsara M (2000) Clinical evaluation of the non-invasive cardiac output (NICO) monitor in the intensive care unit. Anaesth Intensive Care 28(4):427–430 [Comparative Study Evaluation Studies]

77. Nilsson LB, Eldrup N, Berthelsen PG (2001) Lack of agreement between thermodilution and carbon dioxide-rebreathing cardiac output. Acta Anaesthesiol Scand 45(6):680–685

78. Maxwell RA, Gibson JB, Slade JB, Fabian TC, Proctor KG (2001) Noninvasive cardiac output by partial CO2 rebreathing after severe chest trauma. J Trauma 51(5):849–853

79. Valtier B, Cholley BP, Belot JP, de la Coussaye JE, Mateo J, Payen DM (1998) Noninvasive monitoring of cardiac output in critically ill patients using trans-esophageal Doppler. Am J Respir Crit Care Med 158(1):77–83

80. Monchi M, Thebert D, Cariou A, Bellenfant F, Joly LM, Brunet F et al (1998) Clinical evaluation of the Abbott Qvue-OptiQ continuous cardiac output system in critically ill medical patients. J Crit Care 13(2):91–95

81. Burchell SA, Yu M, Takiguchi SA, Ohta RM, Myers SA (1997) Evaluation of a continuous cardiac output and mixed venous oxygen saturation catheter in critically ill surgical patients. Crit Care Med 25(3):388–391

82. Bland JM, Altman DG (1986) Statistical methods for assessing agreement between two methods of clinical measurement. Lancet 1(8476):307–310

83. Hsia CC, Herazo LF, Ramanathan M, Johnson RL Jr (1995) Cardiac output during exercise measured by acetylene rebreathing, thermodilution, and Fick techniques. J Appl Physiol (1985) 78(4):1612–1616

84. Dubin J, Wallerson DC, Cody RJ, Devereux RB (1990) Comparative accuracy of Doppler echocardiographic methods for clinical stroke volume determination. Am Heart J 120(1):116–123

85. Miller WE, Richards KL, Crawford MH (1990) Accuracy of mitral Doppler echocardiographic cardiac output determinations in adults. Am Heart J 119(4):905–910

86. Darmon PL, Hillel Z, Mogtader A, Mindich B, Thys D (1994) Cardiac output by transesophageal echocardiography using continuous-wave Doppler across the aortic valve. Anesthesiology 80(4):796–805; discussion 25A

87. Descorps-Declere A, Smail N, Vigue B, Duranteau J, Mimoz O, Edouard A et al (1996) Transgastric, pulsed Doppler echocardiographic determination of cardiac output. Intensive Care Med 22(1):34–38

88. Cahalan MK, Foster E (1995) Training in trans-esophageal echocardiography: in the lab or on the job? Anesth Analg 81(2):217–218

89. Angelsen BA, Brubakk AO (1976) Transcutaneous measurement of blood flow velocity in the human aorta. Cardiovasc Res 10(3):368–379

90. Huntsman LL, Stewart DK, Barnes SR, Franklin SB, Colocousis JS, Hessel EA (1983) Noninvasive Doppler determination of cardiac output in man. Clinical validation. Circulation 67(3):593–602

91. Kristensen BO, Goldberg SJ (1987) Number of cardiac cycles required to accurately determine mean velocity of blood flow in the ascending aorta and pulmonary trunk. Am J Cardiol 60(8):746–747

92. Singer M (1993) Esophageal Doppler monitoring of aortic blood flow: beat-by-beat cardiac output monitoring. Int Anesthesiol Clin 31(3):99–125

93. Mehta N, Bennett DE (1986) Impaired left ventricular function in acute myocardial infarction assessed by Doppler measurement of ascending aortic blood velocity and maximum acceleration. Am J Cardiol 57(13):1052–1058

94. Abrams JH, Weber RE, Holmen KD (1989) Transtracheal Doppler: a new procedure for continuous cardiac output measurement. Anesthesiology 70(1):134–138

95. Siegel LC, Shafer SL, Martinez GM, Ream AK, Scott JC (1988) Simultaneous measurements of cardiac output by thermodilution, esophageal Doppler, and electrical impedance in anesthetized patients. J Cardiothorac Anesth 2(5):590–595

96. Perrino AC Jr, O'Connor T, Luther M (1994) Transtracheal Doppler cardiac output monitoring: comparison to thermodilution during noncardiac surgery. Anesth Analg 78(6):1060–1066

97. Schmid ER, Spahn DR, Tornic M (1993) Reliability of a new generation transesophageal Doppler device for cardiac output monitoring. Anesth Analg 77(5):971–979

98. Klotz KF, Klingsiek S, Singer M, Wenk H, Eleftheriadis S, Kuppe H et al (1995) Continuous measurement of cardiac output during aortic cross-clamping by the oesophageal Doppler monitor ODM 1. Br J Anaesth 74(6):655–660

99. Spahn DR, Schmid ER, Tornic M, Jenni R, von Segesser L, Turina M et al (1990) Noninvasive versus invasive assessment of cardiac output after cardiac surgery: clinical validation. J Cardiothorac Anesth 4(1):46–59

100. Kubicek WG, Karnegis JN, Patterson RP, Witsoe DA, Mattson RH (1966) Development and evaluation of an impedance cardiac output system. Aerosp Med 37(12):1208–1212

101. Bernstein DP (1986) A new stroke volume equation for thoracic electrical bioimpedance: theory and rationale. Crit Care Med 14(10):904–909

102. Atallah MM, Demain AD (1995) Cardiac output measurement: lack of agreement between thermodilution and thoracic electric bioimpedance in two clinical settings. J Clin Anesth 7(3):182–185 [Comparative Study]

103. Doering L, Lum E, Dracup K, Friedman A (1995) Predictors of between-method differences in cardiac output measurement using thoracic electrical bioimpedance and thermodilution. Crit Care Med 23(10):1667–1673

104. Schoemaker RG, Smits JF (1990) Systolic time intervals as indicators for cardiac function in rat models for heart failure. Eur Heart J 11(Suppl I):114–123 [Research Support, Non-U.S. Gov't]

105. Castor G, Klocke RK, Stoll M, Helms J, Niedermark I (1994) Simultaneous measurement of cardiac output by thermodilution, thoracic electrical bioimpedance and Doppler ultrasound. Br J Anaesth 72(1):133–138 [Comparative Study]

106. Donovan KD, Dobb GJ, Woods WP, Hockings BE (1986) Comparison of transthoracic electrical impedance and thermodilution methods for measuring cardiac output. Crit Care Med 14(12):1038–1044

107. Thomas AN, Ryan J, Doran BR, Pollard BJ (1991) Bioimpedance versus thermodilution cardiac output measurement: the Bomed NCCOM3 after coronary bypass surgery. Intensive Care Med 17(7):383–386

108. Stover JF, Stocker R, Lenherr R, Neff TA, Cottini SR, Zoller B et al (2009) Noninvasive cardiac output and blood pressure monitoring cannot replace an invasive monitoring system in critically ill patients. BMC Anesthesiol 9:6

109. Connors AF Jr, Speroff T, Dawson NV, Thomas C, Harrell FE Jr, Wagner D et al (1996) The effectiveness of right heart catheterization in the initial care of critically ill patients. SUPPORT Investigators. JAMA J Am Med Assoc 276(11):889–897

110. Harvey S, Stevens K, Harrison D, Young D, Brampton W, McCabe C et al (2006) An evaluation of the clinical and cost-effectiveness of pulmonary artery catheters in patient management in intensive care: a systematic review and a randomised controlled trial. Health Technol Assess 10(29):iii–iv, ix–xi, 1–133

111. Lees N, Hamilton M, Rhodes A (2009) Clinical review: goal-directed therapy in high risk surgical patients. Crit Care 13(5):231

112. Funk D, Sebat F, Kumar A (2009) A systems approach to the early recognition and rapid administration of best practice therapy in sepsis and septic shock. Curr Opin Crit Care 15(4):301–307

113. Alhashemi JA, Cecconi M, Hofer CK (2011) Cardiac output monitoring: an integrative perspective. Crit Care 15(2):214

## 3.1 Measurement of Pulmonary Artery Occlusion Pressure by the Pulmonary Artery Catheter

### 3.1.1 Principle

Pulmonary arterial pressure (PAP) is measured at the distal end of the Swan-Ganz catheter. A transient occlusion of blood flow is performed during inflation of the distal balloon in a large caliber pulmonary artery. Beyond the balloon, the pressure drops in the pulmonary artery to a pressure called the pulmonary artery occlusion pressure (PAOP) (Fig. 3.1). This pressure is the same throughout the pulmonary vascular segment in which the balloon is occluded. This segment behaves as an open downstream static column of blood in the pulmonary venous segment. In this regard, the PAOP is a reflection of the pulmonary venous pressure. Because the artery occluded by the balloon is rather large in size, the PAOP is the pressure of a pulmonary vein of the same caliber. Because the resistance of the pulmonary venous segment flowing into the left atrium is considered to be low, the PAOP is a good reflection of the pressure of the left atrium and, by extension, the diastolic pressure of the left ventricle, provided that there is no mitral stenosis. Notably, the PAOP does not match the pulmonary artery wedge pressure. The wedge pressure corresponds to the pressure in relation to the occlusion of a pulmonary vessel of a smaller caliber obtained without

inflating the balloon. Thus, the wedge pressure reflects the pulmonary venous pressure in an area with a lower rating and is greater than the PAOP. Finally, the pulmonary capillary pressure cannot be directly measured. It can only be estimated in two ways, from the decay curve upon balloon inflation or from the Gaar equation, as follows:

$$\text{Pulmonary capillary pressure} = \text{PAOP} + 0.4 \times (\text{PAP}_{\text{mean}} - \text{PAOP})$$

Unfortunately, this formula is only relevant if the venous resistance is homogeneously distributed. Pulmonary capillary pressure is rarely used in clinical practice due to the difficulty of measurement, even though it reliably reflects the risk of pulmonary edema.

### 3.1.2 Validity of the Measurement

It is essential that the intravascular pressure measurement is performed with the utmost care. The reference level during the measurement is the level of the right atrium. This level is between the axillary medium line and the fourth intercostal space. The PAC must be appropriately zeroed and referenced to obtain accurate readings. The choice of zero reference level strongly influences pulmonary pressure readings and pulmonary hypertension classification. One-third of the thoracic diameter best represents the right atrium, while the mid-thoracic level best represents the left

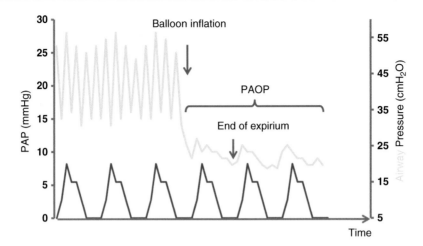

**Fig. 3.1** Measurement of the PAOP from a pulmonary artery catheter in a patient receiving positive-pressure mechanical ventilation (airway pressure curve in *red*)

atrium [1]. Zeroing and referencing should be conducted in one step by always occurring with the patient lying in the recumbent position. However, they represent two separate processes: zeroing involves opening the system to the air to establish the atmospheric pressure as zero, and referencing (or leveling) is accomplished by placing the air-fluid interface of the catheter or transducer at a specific point to negate the effects of the weight of the catheter tubing and fluid column [2]. The system can be referenced by placing the air-fluid interface of either the in-line stopcock or the stopcock that is on top of the transducer at the "phlebostatic level" (i.e., reference point zero). This point is usually the intersection of a frontal plane passing midway between the anterior and posterior surfaces of the chest and a transverse plane lying at the junction of the fourth intercostal space and the sternal margin. Notably, this "phlebostatic level" changes with differences in the position of the patient [3]. This level remains the same regardless of the patient's position in bed (sitting or supine), but it is essential that no lateral rotation occurs. Moreover, it is often difficult to achieve these measures when the patient is in the prone position.

There is a change in intravascular pressure with respiration. During normal spontaneous ventilation, alveolar pressure (relative to atmospheric pressure) decreases during inspiration and increases during expiration. These changes are reversed with positive-pressure ventilation: alveo-

lar pressure increases during inspiration and decreases during expiration. The changes in pleural pressure are transmitted to the cardiac structures and are reflected by changes in pulmonary artery and PAOP measurements during inspiration and expiration.

At end expiration, the pleural and intrathoracic pressures are equal to the atmospheric pressures, regardless of the ventilation mode. Thus, the true transmural pressure and the PAOP should be measured at this point. Transmural pressures at the venous side of both ventricles are known as filling pressures and serve in combination with blood flow as variables for the description of ventricular function. Intrathoracic pressure is not usually available in clinical practice. Therefore, absolute pressures, which depend on transmural pressure, intrathoracic pressure, and the chosen zero level, are used as substitutes.

In healthy patients and patients with spontaneous breathing, the effects of ventilation on intravascular pressures are relatively insignificant. However, these effects are much more pronounced in patients with dyspnea or when the patient is under positive-pressure mechanical ventilation. Therefore, it is imperative that the intravascular pressures are measured at the end of the expiration. At this point, the intrathoracic pressure is closer to the atmospheric pressure. However, if the accessory respiratory muscles are involved in the expiration period, it is necessary to sedate or paralyze the patient or to record these measures at the beginning of the expiration. The intravascular pressure may be overestimated, especially when a positive end-expiratory pressure (PEEP) is applied or in the case of intrinsic PEEP. In these cases, the end-expiratory intrathoracic pressure exceeds the atmospheric pressure. The PEEP values cannot simply be subtracted from the PAOP. Transmission of the alveolar pressure to the intravascular pressure is neither linear nor integral. The presence of lung pathology may affect the coefficient of transmission, e.g., the transmission is attenuated for reduced lung compliance. However, various methods can limit the effects of the PEEP on intravascular pressure, for example, disconnecting the patient from the tube when measuring the PAOP eliminates the influence of the PEEP. Regardless, this method is unsatisfactory because it is accompanied by an increase in the venous return. The PAOP measured off mechanical ventilation does not correspond to the PAOP under positive-pressure ventilation. Another method involves inflating the balloon and then disconnecting the ventilator from the patient. A decrease in PAOP values corresponding to the lowest values of PAOP (nadir PAOP) under mechanical ventila-

tion then occurs in the first 3–4 s after the disconnection [4, 5]. This early measurement taken after disconnection overcomes the venous return. However, disconnection of the tube can cause problems in terms of a loss of alveolar recruitment, particularly in cases of ARDS, and does not solve problems if there is an intrinsic PEEP.

Other authors have proposed a technique based on the fact that PAOP respiratory fluctuations are proportional to respiratory changes in alveolar pressure [6]. It is then possible to calculate the transmission rate corresponding to the difference between the inspiratory and expiratory PAOPs divided by the transpulmonary pressure. This transmission coefficient estimates the alveolar pressure transmission in the intravascular compartment. It is then possible to calculate the PAOP, as corrected according to the following formula:

$$PAOP_{corrected} = \frac{PAOP_{end\,expi} - \left[ PEEP_{total} \times \left( PAOP_{insp} - PAOP_{end\,expi} \right) \right]}{Plateau\,pressure - PEEP_{total}}$$

Using this formula, it is possible to measure the PAOP without disconnecting the ventilator and to account for the intrinsic PEEP (Fig. 3.2).

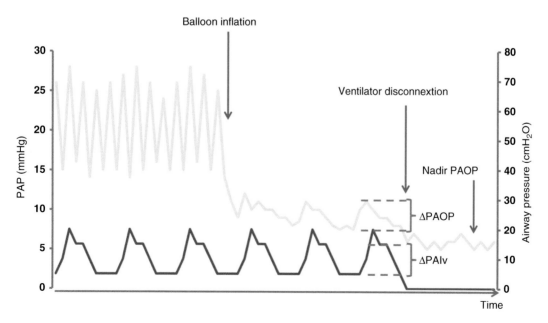

**Fig. 3.2** Measurement of the occluded pulmonary artery pressure (*PAOP*) during ventilation with the PEEP or intrinsic PEEP. When disconnecting the tube, it is possible to measure the "nadir PAOP" and to calculate the trans-

mission of alveolar pressure [6]. ΔPalv represents the plateau pressure – the PEEP – and ΔPAOP is the difference between the peak-inspiratory PAOP and the end-expiratory PAOP

### 3.1.3   Position of the Pulmonary Artery Catheter in the Pulmonary Area

The position of the tip of the pulmonary artery catheter relative to the pulmonary area may affect the validity of PAOP measurements under normal conditions or during application of the PEEP. Lung areas are identified by their relationships among the pressure of the incoming flow (PAP), the pressure of the outgoing flow (pulmonary venous pressure, PvP), and the surrounding pulmonary alveolar pressure (PAlvP) [7] (Fig. 3.3).

Zone I: PAP < PalvP > PvP. Blood does not flow because the pulmonary capillary beds are collapsed. The Swan-Ganz catheter is guided by blood flow, and the tip is usually not moving toward the lung area. The PAOP values are incorrect.

Zone II: PAP > PalvP > PvP. Blood circulates because the blood pressure is greater than the alveolar pressure. Under certain conditions, the catheter tip can be placed in zone II. Measures of the PAOP can be inaccurate.

Zone III: PAP > PAlvP < PvP. The capillaries are open, and blood flows. The tip of the catheter is usually located below the level of the left atrium, and its positioning can be checked by

a lateral thoracic radiograph. Measures of the PAOP are correct.

The distal part of the catheter must be in a lung zone corresponding to zone III, which is the case most of the time because the floating catheter follows the maximum flow. In patients in the supine position, it is positioned in the posterior part, usually on the right side due to the natural curvature of the catheter that is oriented toward the right pulmonary artery. On a chest radiograph, the catheter tip should be located at or below the LA on a plate profile. The PAOP measurement performed in zone II or I would measure the PalvP during inspiration (zone II) or permanently (zone I).

Ventilation, whether spontaneous or controlled, allows a balance of intra- and extra-chest pressure at the end of expiration; measures must be carried out at that time. For example, during inspiration in mechanical ventilation, the catheter area migrates from zone III to zone II. By adding the PEEP, the pulmonary alveolar pressure is increased. By this phenomenon, most of the lungs are found in zone II, inducing a random relationship between the PAOP and LAP. This is particularly noticeable when PEEP values exceed 10 cmH$_2$O. Hypovolemia induces a decrease of the PvP and leads to a passage of the lungs in zone II (Fig. 3.4).

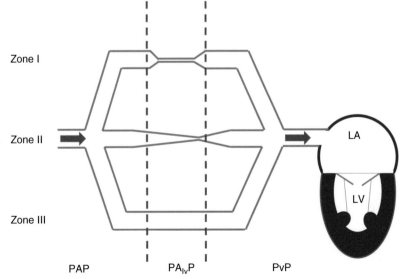

**Fig. 3.3** Schematic lung zones according to JB West and relationships between zones I, II, and III and the pulmonary arterial pressure (*PAP*), pulmonary alveolar pressure (*PAlvP*), and pulmonary venous pressure (*PvP*) [7]. *LA* corresponds to the left atrium, and *LV* corresponds to the left ventricle

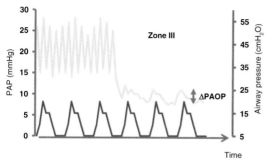
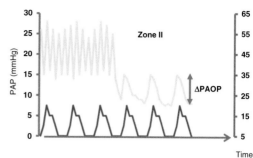
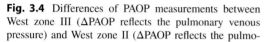

**Fig. 3.4** Differences of PAOP measurements between West zone III (ΔPAOP reflects the pulmonary venous pressure) and West zone II (ΔPAOP reflects the pulmo-nary alveolar pressure) indicating an incorrect position of the pulmonary artery catheter tip

In the case of normal lung compliance, positioning the catheter outside of zone III is recognizable when the PEEP is introduced; the PAOP increases by more than 50 % of the PEEP value and no longer corresponds to the LVEDP values. It is then possible to evaluate the difference by looking at the degree of the PAOP inspiratory rise (Δinsp) compared with the respiratory changes in PAP. If the reported Δinsp PAPO/Δinsp PAP is <1.2, the pulmonary catheter is in zone III, and the measurement of LVEDP is reliable. An inspiratory ratio greater than 1.2 indicates that the PAOP increased in parallel with the PalvP and no longer corresponds to the LVEDP [8]. However, during ARDS, poor lung compliance induces poor pressure transmission, and the model of the West zones is thus not strictly applicable. Taking measurements via a sharp drop in the PEEP leads to obtain values which don't correspond to the actual hemodynamic status. A positive fluid balance with the resulting hypoxemia could be dangerous and may cause an increase in pulmonary arterial resistance. This also applies to patients with COPD because air trapping leads to self-induced PEEP.

The pulmonary vein pressure (PvP) can be pathologically elevated in several situations: fibrosis, mediastinal compression, and thrombosis. Here, Pcap and PAOP are higher than the LA pressure. Reducing the pulmonary vascular bed, e.g., after a pneumonectomy or pulmonary embolism, interrupts the pulmonary flow when occlusion is induced by the balloon, therefore significantly limiting LA filling. In these situations, the PAOP may underestimate the LAP.

### 3.1.4   The Diagnostic Use of Pulmonary Artery Catheter in Circulatory Failure

The hemodynamic profile of a patient can be characterized by measuring intravascular pressure (RAP, PAP, PAOP, and CO). Isolated high PAOP or RAP (CVP) is related to ventricular or valvular dysfunction on the same side. It is important to account for both the absolute value and the ratio between the two pressures [9]. An acute left heart problem, e.g., due to systolic, diastolic, or valvular ventricular dysfunction, is characterized by an isolated elevation of the PAOP. However, it is not possible to differentiate between the two conditions with a pulmonary artery catheter. Hypervolemia or tamponade is suspected when there is a combined increase of the two pressures (CVP and PAOP) [9]. In this case, measuring the cardiac output and the SvO₂ is useful to determine whether hypervolemia (high cardiac output and SvO₂) or tamponade (low cardiac output and low SvO₂) exists.

Right heart dysfunction is suspected in the case of an equalization between the left and right pressures (RAP = PAOP) and if the RAP is greater than the PAOP. Pulmonary hypertension is a sign of an increase in right ventricular afterload (pulmonary embolism, primary or secondary pulmonary hypertension). In contrast, a cardiac pump dysfunction is suspected (ventricular myocardial infarction or tricuspid valve regurgitation) when the PAP is low.

The pulmonary artery catheter is also used to diagnose a pulmonary hypertension and to specify the location and feature. A difference between diastolic mean PAP and PAOP of less than 5 mmHg in the case of pulmonary hypertension is a sign of "postcapillary" PAH (related to an increased left heart pressures). However, if a higher difference between these two pressures exists, then a "precapillary" pulmonary hypertension (primitive pulmonary hypertension, chronic thromboembolic pulmonary hypertension (CTEPH), acute respiratory distress syndrome, pulmonary embolism, decompensated chronic obstructive pulmonary disease) may be suspected. Although these intravascular pressure measurements are diagnostically useful, echocardiography remains essential (impact assessment and possible precision of the exact nature of etiologies). However, the pulmonary artery catheter enables the continuous monitoring of patients in shock.

### 3.1.5   Evaluation of Left Ventricular Preload by the PAOP

To estimate the left ventricular preload, the PAOP must meet a number of criteria:

- The PAOP must be measured in a pulmonary artery with a large enough caliber to reflect the pressure. Indeed, the PAOP corresponds to the pressure of a static column located between the inflated balloon and the pulmonary venous flow, provided that there is no interruption in the pulmonary capillaries. If there is a high alveolar pressure and capillaries are compressed, especially if the intraluminal pressure (i.e., the pulmonary venous pressure) is too low, the pulmonary venous pressure would no longer correspond to the PAOP, which would then be equal to the pulmonary alveolar pressure. To detect such traps, especially in cases of high PEEP (extrinsic or intrinsic), the respiratory changes in the PAOP ($\Delta$PAOP) can be compared with those in the PAP ($\Delta$PAP) [8].

- The left ventricular end-diastolic pressure (LVEDP) should be reflected by the pulmonary vein pressure measured in a larger caliber vein. Indeed, it is very close to the LAP. If there is mitral stenosis, the LAP will be higher than the LVEDP. In this case, the LVEDP will be underestimated by the PAOP. On the other hand, the presence of a "v" wave is the result of acute mitral regurgitation (Fig. 3.5). The PAOP underestimates the LVEDP. In this case, the PAOP must be measured at the beginning of the "v" wave to better estimate the LVEDP.

- It is also important to consider the LVEDP in its "transmural" component to better reflect the LV filling pressure. When there is a high external or

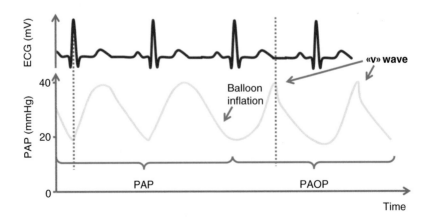

**Fig. 3.5** PAP measurement by a pulmonary artery catheter. During balloon inflation, measurement of the pulmonary arterial occluded pressure (*PAOP*) in the presence of severe mitral insufficiency is reflected on the PAOP curve by a "v" wave. This measure requires the simultaneous monitoring of the pulmonary artery pressure curve and the electrocardiogram (*ECG*)

intrinsic PEEP, including when the LVEDP is reflected by the PAOP, the filling pressure can be overestimated if the PEEP transmitted to the pleural space is not subtracted from the measured PAOP. It is therefore essential to perform this calculation [4, 5]. Finally, in case of reduced left ventricular compliance (e.g., ischemic heart disease or cardiac hypertrophy), the PAOP is not a good reflection of left ventricular volume and preload [10].

### 3.1.6   PAOP as a Marker of Pulmonary Filtration Pressure

The PAOP, as shown above, does not reflect the pulmonary capillary pressure. It is often used to differentiate the type of pulmonary edema (cardiogenic vs. ARDS). In clinical practice, a PAOP above 18 mmHg is often accepted as a sign of the hydrostatic component of pulmonary edema. In this case, the PEEP values are important. Ideally, one should measure a pulmonary capillary pressure that reflects the hydrostatic pressure in the pulmonary capillaries. However, analyzing a decrease in the pulmonary artery pressure curve after balloon inflation is difficult to achieve in clinical practice and is rarely executed. The difference between the pulmonary capillary pressure and the PAOP (pressure measured in a large pulmonary vein) is proportional to the CO and the pulmonary venous resistance. Under physiological conditions, this difference is quite small. However, in some hyperdynamic states such as in ARDS, in which the lung venous resistance is abnormally high, this difference is much greater [11].

### 3.2   Measurement of the Central Venous Pressure via a Central Venous Catheter

### 3.2.1   Central Venous Catheter

The establishment of a central venous catheter (CVC) is a common practice in the ICU. It is essential for the infusion of some drugs such as vasopressors and parenteral nutrients. A CVC also provides the central venous pressure measurements and central venous saturation of the superior vena cava ($ScvO_2$). There are three insertion sites: the internal jugular, subclavian, and femoral veins (long catheters). Although little evidence supports one puncture site over another, each site has its advantages and disadvantages, and the location of the insertion site is made by the clinician, depending on the clinical situation. In patients with shock, the femoral venous route is often selected because of the ease of access and the low risk of pneumothorax. However, the risks of infection and venous thrombosis of the lower limbs, especially for a prolonged catheterization, often lead clinicians to choose a superior vena cava access. Since the first description of an internal jugular CVC insertion was published in 1969 [12], this practice has drastically changed, particularly with the advent of ultrasound-guided techniques. Its insertion by anatomical landmarks is simple, and the catheter route to the superior vena cava is direct. The major disadvantage of this insertion site is the initial puncture of the carotid artery potential pneumothorax, which could be reduced to a negligible risk using ultrasound guidance. The installation success rate of this insertion now exceeds 95 %.

In the ICU, the subclavian route is the most used insertion site. Described for the first time in 1964 [13], this vein has the advantage of being less "collapsible" during profound hypovolemia due to its anatomical grip on the clavicle. Complications occur in 4–15 % of procedures. The risk of pneumothorax ranges from 0 to 6 %. Gas embolisms, arrhythmias, tamponade, and lesions of the nervous structures are extremely rare. As is the case for the internal jugular vein or the femoral vein, ultrasound guidance is recommended for puncturing the subclavian vein. This technique reduces the risk of complications and improves the success of the puncture. Nevertheless, it is always recommended to perform a chest X-ray after the establishment of a CVC. This examination decreases the probability of complications and also checks the proper positioning of the catheter tip.

Catheter infection is the primary risk of complications and occurs in 11 % of cases. Its frequency is dependent on the duration of catheterization. To reduce this risk, training campaigns for nursing staff in hospitals are used [14]. Sterile and aseptic techniques within units have also demonstrated effectiveness. However, the use of catheters coated with antibiotics or antiseptics is still under debate, and tunneling catheters increases infection risks at the femoral and jugular sites [15].

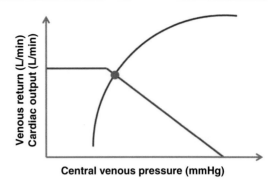

**Fig. 3.6** Cardiac function curve according to the Frank-Starling law (*purple*) and vascular function curve according to Guyton's law (*pink*). The *blue point* represents CVP

## 3.2.2 Central Venous Pressure

The measurement of the central venous pressure (CVP) is carried out via a central venous catheter placed in the lower third of the SVC. The relationship between cardiac output and central venous pressure is twofold: one applies to the heart, and the other applies to the vascular system. The first (the Frank-Starling law) is represented by the cardiac function curve. Cardiac output varies with preload, as expressed by the CVP. The main mechanisms that govern this function are afterload and contractility. The second mechanism concerns the vascular function, for which the CVP varies inversely with the cardiac output according to the Guyton vascular function curve law [16]. The main determinants of vascular function are the arterial and venous compliances, the peripheral vascular resistance, and the blood volume. The intersection of the cardiac and vascular function curves reflects a state of equilibrium (Fig. 3.6).

The main question asked by the intensivist at the bedside of a patient with shock is whether volume expansion will be beneficial [17]. Until the early 2000s, estimated volemia, representing the total blood volume in the body, interested both clinicians and researchers. Its determination is difficult in the ICU and has little practical significance because it is only an indicator of a patient's volume status and not blood circulation. The assessment of preload is also a key element to consider. It roughly corresponds to the ventricular loading conditions at the end-diastolic time. The relationship between the preload and the stroke volume

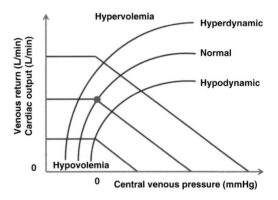

**Fig. 3.7** Graphic representation of the inseparable combination of the curves of right ventricular function and venous return to different hemodynamic states. The represented intersections symbolize the different states of cardiac function; the *blue dot* represents the steady-state condition

can distinguish between two types of patients and helps to define the hemodynamic response to fluid expansion. The "responder" patient (i.e., "preload dependent") is a patient in whom a volume expansion will lead to a significantly increased SV and, accordingly, CO (for a small increase in the transmural pressure). This patient will be situated on the vertical portion of the cardiac function curve. The "nonresponder" patient (i.e., "pre-independent") is a patient in whom a volume expansion will lead to an increased preload due to an increased transmural pressure but no significant increase in the stroke volume. This patient will be located on the plateau portion of the cardiac function curve (Fig. 3.7).

Regarding hypovolemia, we must distinguish between absolute hypovolemia and relative hypovolemia. Absolute hypovolemia indicates a decrease in the total circulating blood volume. It results in a decrease of the systemic venous return, the cardiac preload, and, thus, the cardiac output, despite a reactive increase in the heart rate. The relative hypovolemia is defined by inadequate blood volume distribution between different compartments, as blood volume may be defined as stressed and unstressed blood volume. This results in a decrease in the central blood volume corresponding to the intrathoracic blood volume and is especially the case during positive-pressure mechanical ventilation or vein dilatation (decrease in stressed blood volume in favor of unstressed blood volume). Concerning the CVP, as shown in the various states in Fig. 3.7, a same venous return curve corresponds to several CVP values, depending finally from the good cardiac function or the impaired heart function. Therefore, analyzing the CVP according to the CO is essential. The Frank-Starling relationship may vary from one patient to another and over time in the same patient.

Under the Frank-Starling law governing the relationship between preload and ventricular function, there are two phases on the cardiac function curve. During the rising phase, the increased preload results in an increase in stroke volume. In the plateau phase, an increase in the preload does not cause an increase in the stroke volume. In contrast, the plateau represents the filling limit of the ventricles in connection with external components such as the pericardium and the cytoskeleton. On this portion of the curve, an increase in the preload increases the diastolic ventricular pressure and the left ventricular transmural pressure with negative potential consequences on the coronary circulation of the left ventricular, hepatic, and renal flows. Finally, venous collapse can occur and limit venous return [18].

However, it is more complicated to estimate the "driving" pressure at the periphery of the veins. In fact, the venous pressure is variable throughout the body, particularly when the patient is in an orthostatic position, due to the weight of the blood column itself. These variations are more important in the supine position because the height between the front and the rear body rarely exceeds 30 cm. In a study on dogs deprived of sympathetic reflexes and with hearts replaced by pumps, Guyton measured the "mean" driving pressure of venous return or the mean systemic pressure (MSP). In this experiment, increasing the pressure of the right atrium to more than 7 mmHg nullified the venous return and cardiac output. This indicated that the atrial pressure reached the MSP value and thus canceled out the venous return motor gradient [16, 19]. Therefore, the venous return in these dogs occurred with a maximum gradient of 7 mmHg. The present fact is only possible because the venous system offers little resistance to flow, unlike the arterial network.

If the normal pressure of the right atrium is close to 0 mmHg, it is not uncommon to measure a RAP ≥7 mmHg in patients under positive-pressure ventilation or suffering from impaired right ventricular function (without venous return). As a result, the cardiac output is not zero even if it may be significantly reduced. This is related to a parallel increase in the MSP due to a reflexive increase of venoconstrictor tone. Conversely, decreasing the RAP below 0 mmHg may not increase the venous return due to the collapse of the vena cava when the transmural pressure is zero or negative (resulting in no flow) [16, 20]. This is shown as a plateau in the venous return curve when the inferior vena cava collapses at the level of the diaphragm (abdominal pressure is higher than intrathoracic pressure) (Fig. 3.8).

### 3.2.3 Measurement of the Mean Systemic Pressure

It is relatively simple to measure the level of right atrium pressure that opposes the venous return.

### 3.2.4 Resistance to Venous Return

The resistance to venous return is very low, but minor changes can have major consequences in terms of flow because the pressure gradient is

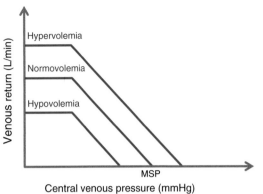 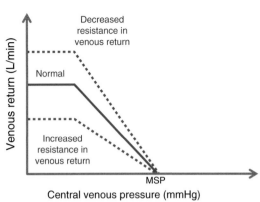

**Fig. 3.8** Venous return curves as described by Guyton [16]. For some CVP values, the venous return is canceled out. In contrast, for values over 0 mmHg, the flow does not increase due to the collapse of the vena cava where the transmural pressure becomes negative. In addition, the slope of the curve is inversely proportional to the resistance to venous return

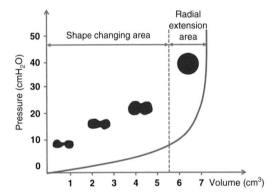

**Fig. 3.9** The pressure-volume relationship of a canine jugular vein indicating the shape-changing area and the radial extension area. The shape-changing area coincides with the preload-dependent area

also very low. Cylindrical veins offer low resistance. However, for flattened or collapsed veins, the resistance increases and becomes infinite (Fig. 3.9). The venous return curve slope is the inverse of the venous return resistance: for the same MSP value, a steep slope, indicating a low resistance, allows a greater venous return.

### 3.2.5   Venous Reservoir and Cardiac Output

The venous reservoir can be represented as a container with a port located above a bottom portion [21, 22]. The contained liquid can therefore be divided into a portion located below the level of the port, corresponding to an "unstressed volume," and a portion located above the port, corresponding to a "stressed volume." The fraction of unstressed blood volume is passively stored in the veins and can be used without producing distending pressure [23]. This is the volume that is used to "prime" the circuit but that generates no flow. The stressed volume is located above the port. The higher the liquid above the level of the orifice, the greater the hydrostatic pressure and, therefore, the greater the venous return and the CO. This height is the driving pressure gradient of the venous return and is equivalent to the difference between the MSP and the RAP. Thus, to increase the venous return, it is possible either to increase the MSP or to lower the RAP (Fig. 3.10). To increase the MSP, two methods can be used: (a) increasing the volume in the reservoir (e.g., volume expansion) and (b) reducing the capacitance of the reservoir by administering a vasoconstrictor agent (to redistribute the volumes by increasing the stressed volume at the expense of the unstressed volume). To reduce the RAP without reducing the MSP, an inotropic agent can be administered to increase the contractility of the ventricles and decrease the amount of fluid in the upstream atrium. Conversely, a decrease in the volume of the reservoir, e.g., through bleeding or dehydration, will have the effect of reducing the venous return and the CO.

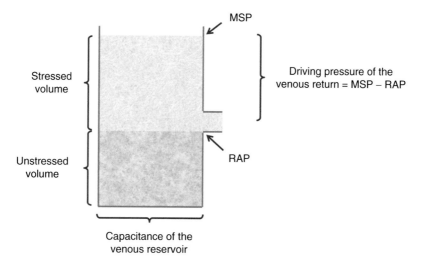

**Fig. 3.10** Schematic representation of the venous reservoir. The size of the container is the capacitance of the reservoir, which is at a maximum when veins are dilated. The height of the orifice corresponds to the RAP. The total liquid height corresponds to the MSP. The volume of liquid located below the level of the orifice corresponds to the unstressed volume generating no flow, whereas the volume located above corresponds to the stressed volume. The liquid height located above the orifice corresponds to the driving pressure of the venous return, i.e., the difference between the MSP and the RAP

Changes in the intravascular volume and the venous capacitance affect the MSP and the venous return resistance [24]. Therefore, fluid expansion or venoconstriction induces an increase of the venous return by increasing the MSP and decreasing the resistance in the venous return by recruiting collapsed or flattened veins. Dehydration or hemorrhage results in the opposite effect (Fig. 3.11).

### 3.2.6   CVP Measurement Principles

From a physiological point of view, CVP measurement must take into account two properties: the reference value and physiological variations. For each measure, it is necessary to have a corresponding reference value. This is most often an arbitrary value because different values of CVP will be obtained for each baseline. For example, the CVP measured at the midaxillary level will be greater by 3 mmHg than that measured at the sternal angle. The implementation of a "zero" reference is required before each measurement. CVP measurements are carried out in the vast majority of cases through a central venous line located at the superior vena cava. However, in the absence of an abdominal compartment syndrome, measuring the CVP in the inferior vena cava is feasible. These two sites of measurements were compared in clinical studies and showed good correlation [25]. However, in these studies, the tip of the venous catheter was consistently located above the diaphragm, which may not always be the case in clinical practice. This is a major limitation of CVP measurements by a femoral catheter. Similarly, studies have compared CVP values (with good correlation) measured centrally vs. peripherally in renal transplant patients with no history of heart disease during and after surgery [26, 27]. Nonetheless, performing these measurements in clinical practice is not recommended due to the lack of reliable data and other clinical factors that may distort the measured values.

The interaction between the ventilation and the CVP curve through the transmural pressure is the cause of variations in CVP curves. In a patient with spontaneous breathing, forced inspiration induces a reduction in the CVP. In contrast, in a patient under mechanical ventilation with positive pressure, the "zero" reference is equal to the atmospheric pressure. During

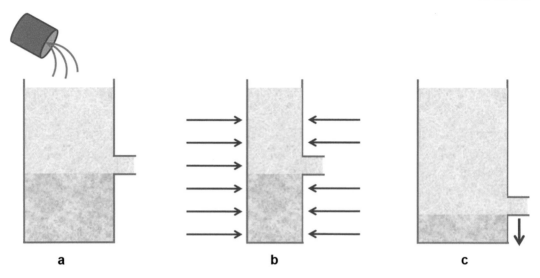

**Fig. 3.11** Schematic representation of three ways to increase the venous return and the CO: (**a**) by increasing MSP through a volume expansion, (**b**) by increasing the MSP by administering a vasoconstrictor, (**c**) by lowering the RAP by administering an inotropic agent

mechanical ventilation with positive pressure, the CVP value increases as a result of the sharp increase in the surrounding pressure of heart and vessels (extramural pressures), the fact that decreases the transmural pressure and the size of the right atrium. However, large differences are still observed, especially during the application of positive-pressure ventilation in the case of abdominal compartment syndrome or in the presence of pericardial effusion. No solutions have been proposed for the reliable and reproducible measurement of CVP values [18] under unphysiological conditions.

### 3.2.6.1 Measurement of CVP

The CVP curve comprises several waves: three ascending deflections (a, c, and v) and two descending waveforms (x and y). The "a" waveform is due to contraction of the right atrium subsequent to the electrical stimulation and P wave of the ECG. The "c" wave is attributed to the isovolumetric contraction of the right ventricle that induces a bulging tricuspid valve toward the right atrium. The "x" wave is attributed to decreased pressure in the right atrium, which opens the tricuspid valve to the bottom during ejection of the right ventricle. The "v" wave is formed by the opening of the tricuspid valve as blood enters the

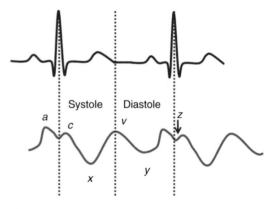

**Fig. 3.12** Electroscopic trace of the central venous pressure curve. The optimum measurement is achieved at the point "Z"

right ventricle. Point "z" is the atrial pressure before ventricular contraction (Fig. 3.12). There are approximately 200 ms between the CVP curve and the radial arterial pressure curve. Therefore, there is an "artificial" delay between systole transmitted by the radial artery and the systolic "c" wave of the CVP.

### 3.2.6.2 How to Use CVP Measurements in Clinical Practice

The measurement of CVP values is used to estimate the pressure in the right atrium. This reflects the right ventricle diastolic pressure, which estimates the diastolic volume of the right ventricle. Finally, the CVP can be used as a surrogate to estimate the right ventricular preload, as it is an indicator of the interaction between venous return and right ventricular function [28]. Clinicians have used the CVP as an indicator of volemia. Although the CVP varies with volume in healthy subjects, for instance, studies have shown that its measure is unnecessary in patients with heart failure, especially if the left ventricular ejection fraction is decreased [29]. Moreover, the CVP has no predictive value for fluid responsiveness [30, 31]: it does not distinguish responders from nonresponders to volume expansion. Finally, CVP values in patients under positive-pressure ventilation [32] or with abdominal compartment syndrome should be interpreted with caution [33]. In particular, PEEP may influence the CVP. A simple subtraction does not determine the actual CVP value. However, a high CVP value is often notably present in the case of right heart failure such as in pulmonary embolism [34] and should be considered a warning sign to the clinician. Higher values of the CVP also predict the occurrence of right heart failure in the establishment of left ventricular assistance. One study showed that an important rise in the CVP during the implantation of a left ventricular assist device predicts the occurrence of right ventricular dysfunction.

Low CVP values can still assist the clinician in treatment decisions, especially in cases of hypovolemic shock, in severe trauma patients, and during some perioperative surgeries. This is especially relevant in emergency services for which the patient is breathing spontaneously without positive-pressure ventilation or deep sedation and has an irregular heart rate as in these conditions, dynamic indices of fluid responsiveness are useless. Accordingly, Rivers et al. established their early management protocol for patients in septic shock, in which the CVP takes precedence in the initial treatment strategy [35]. For example, for a CVP <8 mmHg, the clinician is advised to achieve volume expansion. These practices were adapted by the Surviving Sepsis Campaign [36]. Although CVP measurement should not be the only index considered, it could be a primary factor among others in the overall treatment process.

## References

1. Kovacs G, Avian A, Olschewski A, Olschewski H (2013) Zero reference level for right heart catheterisation. Eur Respir J 42(6):1586–1594
2. Summerhill EM, Baram M (2005) Principles of pulmonary artery catheterization in the critically ill. Lung 183(3):209–219
3. Bridges EJ, Woods SL (1993) Pulmonary artery pressure measurement: state of the art. Heart Lung 22(2):99–111
4. Carter RS, Snyder JV, Pinsky MR (1985) LV filling pressure during PEEP measured by nadir wedge pressure after airway disconnection. Am J Physiol 249(4 Pt 2):H770-6. Research Support, Non-U.S. Gov't Research Support, U.S. Gov't, Non-P.H.S
5. Pinsky M, Vincent JL, De Smet J (1991) Estimating left ventricular filling pressure during positive end-expiratory pressure in humans. Am Rev Respir Dis 143(1):25–31. Research Support, Non-U.S. Gov't
6. Teboul JL, Pinsky MR, Mercat A, Anguel N, Bernardin G, Achard JM et al (2000) Estimating cardiac filling pressure in mechanically ventilated patients with hyperinflation. Crit Care Med 28(11):3631–3636
7. West JB, Dollery CT, Naimark A (1964) Distribution of blood flow in isolated lung; relation to vascular and alveolar pressures. J Appl Physiol 19:713–724
8. Teboul JL, Besbes M, Andrivet P, Axler O, Douguet D, Zelter M et al (1992) A bedside index assessing the reliability of pulmonary artery occlusion pressure measurements during mechanical ventilation with positive end-expiratory pressure. J Crit Care 7(1):22–29
9. Jones JW, Izzat NN, Rashad MN, Thornby JI, McLean TR, Svensson LG et al (1992) Usefulness of right ventricular indices in early diagnosis of cardiac tamponade. Ann Thorac Surg 54(1):44–49
10. Crexells C, Chatterjee K, Forrester JS, Dikshit K, Swan HJ (1973) Optimal level of filling pressure in the left side of the heart in acute myocardial infarction. N Engl J Med 289(24):1263–1266
11. Her C, Mandy S, Bairamian M (2005) Increased pulmonary venous resistance contributes to increased pulmonary artery diastolic-pulmonary wedge pressure gradient in acute respiratory distress syndrome. Anesthesiology 102(3):574–580. Research Support, Non-U.S. Gov't
12. English IC, Frew RM, Pigott JF, Zaki M (1969) Percutaneous cannulation of the internal jugular vein. Thorax 24(4):496–497
13. Baden H (1964) Percutaneous catheterization of the subclavian vein. Nord Med 71:590–593
14. Zingg W, Cartier V, Inan C, Touveneau S, Theriault M, Gayet-Ageron A et al (2014) Hospital-wide

multidisciplinary, multimodal intervention programme to reduce central venous catheter-associated bloodstream infection. PLoS One 9(4):e93898

15. Aitken EL, Stevenson KS, Gingell-Littlejohn M, Aitken M, Clancy M, Kingsmore DB (2014) The use of tunneled central venous catheters: inevitable or system failure? J Vasc Access 0(0):0

16. Guyton AC, Lindsey AW, Abernathy B, Richardson T (1957) Venous return at various right atrial pressures and the normal venous return curve. Am J Physiol 189(3):609–615

17. Guerin L, Monnet X, Teboul JL (2013) Monitoring volume and fluid responsiveness: from static to dynamic indicators. Best Pract Res Clin Anaesthesiol 27(2):177–185

18. Magder S (2005) How to use central venous pressure measurements. Curr Opin Crit Care 11(3):264–270

19. Guyton AC, Richardson TQ, Langston JB (1964) Regulation of cardiac output and venous return. Clin Anesth 3:1–34

20. Guyton AC, Adkins LH (1954) Quantitative aspects of the collapse factor in relation to venous return. Am J Physiol 177(3):523–527

21. Bressack MA, Raffin TA (1987) Importance of venous return, venous resistance, and mean circulatory pressure in the physiology and management of shock. Chest 92(5):906–912

22. Sylvester JT, Goldberg HS, Permutt S (1983) The role of the vasculature in the regulation of cardiac output. Clin Chest Med 4(2):111–126

23. Magder S, De Varennes B (1998) Clinical death and the measurement of stressed vascular volume. Crit Care Med 26(6):1061–1064

24. Bressack MA, Morton NS, Hortop J (1987) Group B streptococcal sepsis in the piglet: effects of fluid therapy on venous return, organ edema, and organ blood flow. Circ Res 61(5):659–669

25. Dillon PJ, Columb MO, Hume DD (2001) Comparison of superior vena caval and femoroiliac venous pressure measurements during normal and inverse ratio ventilation. Crit Care Med 29(1):37–39

26. Amar D, Melendez JA, Zhang H, Dobres C, Leung DH, Padilla RE (2001) Correlation of peripheral venous pressure and central venous pressure in surgical patients. J Cardiothorac Vasc Anesth 15(1):40–43

27. Hadimioglu N, Ertug Z, Yegin A, Sanli S, Gurkan A, Demirbas A (2006) Correlation of peripheral venous pressure and central venous pressure in kidney recipients. Transplant Proc 38(2):440–442

28. Osman D, Monnet X, Castelain V, Anguel N, Warszawski J, Teboul JL et al (2009) Incidence and prognostic value of right ventricular failure in acute respiratory distress syndrome. Intensive Care Med 35(1):69–76

29. Mangano DT (1980) Monitoring pulmonary arterial pressure in coronary-artery disease. Anesthesiology 53(5):364–370

30. Boulain T, Achard JM, Teboul JL, Richard C, Perrotin D, Ginies G (2002) Changes in BP induced by passive leg raising predict response to fluid loading in critically ill patients. Chest 121(4):1245–1252

31. Michard F, Teboul JL (2002) Predicting fluid responsiveness in ICU patients: a critical analysis of the evidence. Chest 121(6):2000–2008

32. Chen FH (1985) Hemodynamic effects of positive pressure ventilation: vena caval pressure in patients without injuries to the inferior vena cava. J Trauma 25(4):347–349

33. Cheatham ML (2009) Abdominal compartment syndrome: pathophysiology and definitions. Scand J Trauma Resusc Emerg Med 17:10

34. Cheriex EC, Sreeram N, Eussen YF, Pieters FA, Wellens HJ (1994) Cross sectional Doppler echocardiography as the initial technique for the diagnosis of acute pulmonary embolism. Br Heart J 72(1):52–57

35. Rivers E, Nguyen B, Havstad S, Ressler J, Muzzin A, Knoblich B et al (2001) Early goal-directed therapy in the treatment of severe sepsis and septic shock. N Engl J Med 345(19):1368–1377. Clinical Trial Randomized Controlled Trial Research Support, Non-U.S. Gov't

36. Dellinger RP, Levy MM, Rhodes A, Annane D, Gerlach H, Opal SM et al (2013) Surviving Sepsis Campaign: international guidelines for management of severe sepsis and septic shock, 2012. Intensive Care Med 39(2):165–228. Practice Guideline

# Monitoring the Adequacy of Oxygen Supply and Demand

## 4.1 Physiological Basis

One of the main goals of blood circulation is to ensure oxygen supply to organs and tissues. The determinants of arterial oxygen delivery ($DO_2$) are the CO and the arterial oxygen content ($CaO_2$). The arterial oxygen content has two components; the main component is oxygen bound to hemoglobin ($SaO_2$), and the secondary component is dissolved oxygen. The former is dependent on the hemoglobin concentration; the affinity of hemoglobin for oxygen (which varies for Hb isotypes); environmental conditions such as temperature, pH, or 2,3-DPG concentrations; and thus the Hb oxygen saturation. The second component is dependent on the arterial partial pressure of oxygen ($PaO_2$) and is considered to be negligible due to the very low solubility coefficient of oxygen in plasma (close to 0). It is therefore possible to set the equations:

$$CaO_2 = \left(Hb \times 1.34 \times SaO_2\right) + \left(0.003 \times PaO_2\right)$$

$$DO_2 = CO \times CaO_2$$

By ignoring the dissolved oxygen component, we obtain:

$$DO_2 = CO \times Hb \times 1.34 \times SaO_2$$

Arterial blood is normally deoxygenated in tissues. Tissue oxygen extraction is dependent on tissue demand but also on the ability of the tissue to extract oxygen. Therefore, following peripheral oxygen extraction, the venous oxygen content is dependent on the arterial oxygen saturation ($SaO_2$), on the balance between $VO_2$ and the cardiac output (CO), and on hemoglobin (Hb) levels.

As a surrogate of $SvO_2$ for evaluating the adequacy of $O_2$ supply/demand, the central oxygen venous saturation ($ScvO_2$) has become a commonly used variable. Because it represents the amount of oxygen remaining in the systemic circulation after its passage through the tissues, the $ScvO_2$ informs us of the balance between oxygen transport ($DO_2$) and oxygen consumption ($VO_2$). Its use in clinical practice was facilitated over a decade ago by the availability of fiber optic catheters that allow continuous monitoring [1]. A reduction in the cardiac output, in hemoglobinemia, or in the $SaO_2$ or an excessive $VO_2$ may initially be compensated for by an increase in the arteriovenous oxygen difference, resulting in a decreased $ScvO_2$. This is an early compensatory mechanism that can precede a rise in lactatemia [2]. $ScvO_2$ values of <65–70 % under acute patient conditions should alert clinicians to the presence of tissue hypoxia or inadequate perfusion.

## 4.2 Mixed Venous Oxygen Saturation ($SvO_2$)

The pulmonary artery catheter permits measurement of the mixed $SvO_2$. There are two ways to achieve this:

R. Giraud, K. Bendjelid, *Hemodynamic Monitoring in the ICU*, DOI 10.1007/978-3-319-29430-8_4

1. A sample of blood is taken from the pulmonary artery through the distal port of the pulmonary artery catheter (balloon deflated) and subjected to conventional blood gas measurements by co-oximetry. However, this method has multiple potential pitfalls that should be avoided during the removal of pulmonary arterial blood [3]. Strict sampling rules must be followed to prevent the collection of non-arterialized mixed venous blood. The correct positioning of the catheter tip in a large branch of the pulmonary artery is essential. The measuring method by co-oximetry has also been a source of frequent errors. This method also may potentially cause major blood loss, especially in younger children, and is also the source of infections associated with frequent handling of the pulmonary artery catheter.

2. A pulmonary artery catheter fitted with an optical fiber is used for the in vivo measurement and continuous recording of the $SvO_2$ via automatic spectrophotometry. This method avoids repeated pulmonary arterial sampling. It also allows real-time $SvO_2$ monitoring. This method is very accurate and reproducible and uses several wavelengths [3]. The measuring principle is based on red and infrared light sources that send 600–1,000 nm wavelengths through the optical fiber of the pulmonary artery catheter to illuminate the blood flow from the pulmonary artery. The reflected light is captured by a photodetector through a second optical fiber. These captured readings are then integrated to determine the $SvO_2$. An "in vitro" calibration must be conducted before insertion of the pulmonary artery catheter. Once the catheter is in place, a supplementary "in vivo" calibration, in which a pulmonary artery blood sample is measured, may be performed. It is also recommended that the calibration be repeated when the $SvO_2$ values are suspicious or erroneous. The position of the catheter in the pulmonary artery (i.e., not too distally) is the main factor that determines the precision of the measured $SvO_2$. Manufacturers claim measurement precisions of ±2%. However, in a study comparing this method with co-oximetry, the average precision varied

by up to 9%. In clinical practice, −9% to +9% variations are acceptable [4]. These variations are often due to poor positioning of the catheter or improper use of the device rather than to a poor-quality device [5]. Once properly repositioned and recalibrated, the pulmonary artery catheter measurement system often reduces erroneous $SvO_2$ values.

$SvO_2$ measurements assess the adequacy of oxygen delivery ($DO_2$) and oxygen consumption ($VO_2$). $SvO_2$ is affected in part by the cardiac output, the arterial oxygen saturation ($SaO_2$), the hemoglobin concentration (Hb), and the $VO_2$. Based on the Fick relationship, the $SvO_2$ can be calculated using the following equation:

$$SvO_2 = SaO_2 - \frac{VO_2}{\dot{Q} \times Hb \times 13.9}$$

The relationship between the cardiac output and the $SvO_2$ is curvilinear [6] (Fig. 4.1) for given $SaO_2$, $VO_2$, and Hb values. A low $SvO_2$ is associated with a decreased cardiac output. In contrast, a normal $SvO_2$ (≥70%) is associated with a normal or increased cardiac output. Additionally, when the $SvO_2$ is low, any changes in the cardiac output are associated with changes in the $SvO_2$. However, for normal or high $SvO_2$ values (>70%), significant changes in the cardiac output are associated with small changes in the $SvO_2$. Therefore, a decoupling phenomenon exists between the cardiac output and the $SvO_2$. This precludes the use of this single monitoring system to assess cardiac output changes, especially during hyperdynamic states.

In healthy subjects at rest with normal $SaO_2$ and Hb values, the normal $SvO_2$ value is 70–75%. During exercise, $SvO_2$ values may decrease to as low as 45% [7], due to an increase in $O_2$ consumption, with both increase in $VO_2$ and $O_2$ extraction by skeletal muscle. However, anaerobic metabolism occurs at this "critical" $SvO_2$, which also corresponds to the $O_2$ tissue extraction limit (or critical extraction). In certain pathological situations, the drop in the $SvO_2$ is the result of complex interactions between four determinants that could all be influenced to varying degrees by pathology or therapy. The

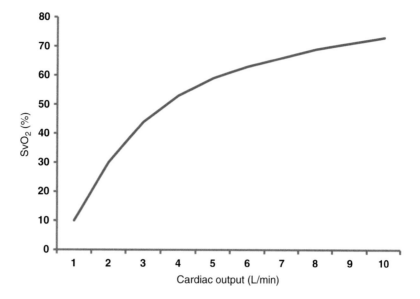

**Fig. 4.1** Relationship between the $SvO_2$ and the cardiac output. The $SvO_2$/CO relationship is curvilinear, with constant Hb, $SaO_2$, and $VO_2$ values

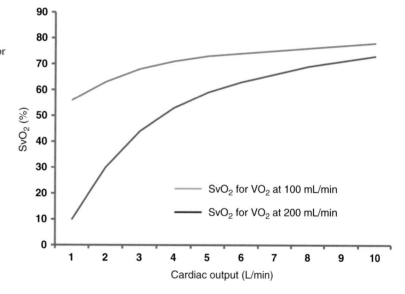

**Fig. 4.2** Relationship between $SvO_2$ and cardiac output. The $SvO_2$/CO relationship is curvilinear. For constant Hb, $SaO_2$, and $VO_2$ values, CO variations cause large $SvO_2$ variations when the initial CO value is low. Conversely, for high CO values, variations do not affect $SvO_2$ values. These relationships are changed when changes to the CO are accompanied by changes in the $VO_2$

four determinants $SaO_2$, CO, Hb, and $VO_2$ are closely linked through various compensatory mechanisms (Fig. 4.2).

## 4.3 SvO₂ and Regional Oxygenation

The $SvO_2$ is measured by a pulmonary artery catheter and is a reflection of the average saturation of venous blood in organs. A few organs such as the kidneys have high blood flow perfusion and correspondingly low $O_2$ extraction. These organs have a greater influence on $SvO_2$ values than other organs such as the myocardium that are perfused at lower flow rates and with greater $O_2$ extraction. During sepsis, there is a disturbance in the blood flow distribution between organs, which complicates the understanding of the measured value of the $SvO_2$. This is particularly true in the hepato-splanchnic compartment, where there is a poor distribution of regional flow in septic shock and which is associated with higher $O_2$ consumption [8]. In the present setting, hypoperfusion and dysoxia, present in the splanchnic region, are partly responsible for multiple-organ failure [9]. Another example of the present phenomenon is the demonstration in some patients with septic shock of a normal $SvO_2$ value while very low values of $O_2$ saturation at the level of the hepatic veins are observed [10, 11]. Therefore, it appears that the $SvO_2$ is not a reliable monitor of regional perfusion in some kind of shocks like circulatory failure related to sepsis.

## 4.4   Contributions of ScvO$_2$

Whereas the $SvO_2$ reflects the venous oxygenation of the whole body and requires the presence of a pulmonary artery catheter, the $ScvO_2$ is a reflection of the venous oxygenation of the brain and the upper body. Its measurement is possible through a central venous catheter placed in the superior vena cava at the level of the right atrium. The mixed $SvO_2$ is a mixture of venous blood from the inferior vena cava territories, the superior vena cava, and the coronary sinus. However, the $SvO_2$ is dependent on each organ because each organ extracts different amounts of $O_2$. Under normal physiological conditions, the $SvO_2$ is higher in the lower body than in the upper body [12, 13]. Under certain pathological conditions, this difference is reversed [14]. During general anesthesia, due to the increase in cerebral blood flow and the use of anesthetic drugs that induce a reduction in brain $O_2$ extraction, the $ScvO_2$ is often greater than the $SvO_2$ by approximately 5 % [15]. A similar effect is observed in severe head trauma patients treated with barbiturates. In shock, mesenteric blood flow decreases, whereas $O_2$ extraction increases in the same region. In contrast, the $ScvO_2$ increases in the region of the superior vena cava because blood flow is maintained. Therefore, the venous saturation of the inferior vena cava decreases, and the $SvO_2$ may be lower than the $ScvO_2$ [16].

However, the question remains whether the two venous saturations are equivalent, interchangeable, or move in the same direction during pathological situations. Numerous studies in humans and in animals have shown contradictory results. A few studies have reported surprisingly similar values [2, 17, 18], though others have reported significantly different values [19, 20].

The trend of the past 10 years has been to use less invasive monitoring techniques and to shift from measuring the $SvO_2$ to the $ScvO_2$. Moreover, Rivers et al. conducted a randomized study based on the early management of patients with septic shock. The objective was to evaluate the efficacy of a protocol based on early therapeutic goals, especially one wherein the $ScvO_2$ values had to be greater than or equal to 70 % during the first 6 h of care. These protocols were based on volume expansion, catecholamine administration and packed red cell transfusion. The results of this study showed that the relative risk of death at 60 days in the group treated with this protocol significantly improved compared with a conventionally treated group [21]. Although these results have been questioned on numerous occasions, this study has shown the advantages of the early and aggressive management of septic patients based on the monitoring of an easily accessible oxygenation criterion. Since then, the relevance of this parameter for improving the prognosis of patients in shock has been shown by many other studies conducted in the ICU [22, 23].

Nevertheless, it is important at this stage to define the limits of the $SvO_2$ and $ScvO_2$ values during sepsis. First, one could argue that ScvO2 measurement requires a central venous catheter, which is an invasive technique that exposes patients to complications such as infection or hemorrhage. However, central venous lines are often required in critical patients and could therefore be used for ScvO2 monitoring. Although catheter placement has been a subject of debate, good correlation and parallelism have been observed between mixed venous blood saturation and the ScvO2 in critical patients over a broad range of clinical situations [24]. Second, given its ability to measure global DO2, the ScvO2 is unable to assess local perfusion deficits [25, 26]. Consequently, in situations for which the microcirculation is greatly altered (e.g., sepsis and late-phase shock states) or in mitochondrial poisoning or dysfunction, the ScvO2 may present increased values coexisting with situations of intense tissue hypoxia [27].

To conclude about the $ScvO_2$, the presence of ScvO2 <60 % in the general critical patient population is associated with increased mortality [28]. ScvO2 measurement, as one of predefined resuscitation goals, appears to be a valuable tool in the early phase of septic shock (before volume resuscitation) in guiding fluid management and inotrope support. Nevertheless, a greater knowledge of its determinants is essential to ensure a reliable interpretation in clinical practice. When the ScvO2 is low, it reflects an unbalance between oxygen consumption and oxygen supply and should lead to the proposal of an appropriate optimization strategy. Additionally, in clinical situations such as septic shock, after the first hours of management, a "normal" or high ScvO2 provides no additional value. Despite the extent and the limits of $ScvO_2$ interpretation, $ScvO_2$ monitoring is now an integral part of management algorithms such as the Surviving Sepsis Campaign [29], though some recent studies have shown that early goal-directed therapy protocol did not lead to improved outcomes [30, 31].

# References

1. Scalea TM, Hartnett RW, Duncan AO, Atweh NA, Phillips TF, Sclafani SJ et al (1990) Central venous oxygen saturation: a useful clinical tool in trauma patients. J Trauma 30(12):1539–1543
2. Berridge JC (1992) Influence of cardiac output on the correlation between mixed venous and central venous oxygen saturation. Br J Anaesth 69(4):409–410
3. Cariou A, Monchi M, Dhainaut JF (1998) Continuous cardiac output and mixed venous oxygen saturation monitoring. J Crit Care 13(4):198–213
4. Scuderi PE, Bowton DL, Meredith JW, Harris LC, Evans JB, Anderson RL (1992) A comparison of three pulmonary artery oximetry catheters in intensive care unit patients. Chest 102(3):896–905
5. Kim KM, Ko JS, Gwak MS, Kim GS, Cho HS (2013) Comparison of mixed venous oxygen saturation after in vitro calibration of pulmonary artery catheter with that of pulmonary arterial blood in patients undergoing living donor liver transplantation. Transplant Proc 45(5):1916–1919
6. Giraud R, Siegenthaler N, Gayet-Ageron A, Combescure C, Romand JA, Bendjelid K (2011) ScvO(2) as a marker to define fluid responsiveness. J Trauma 70(4):802–807
7. Weber KT, Andrews V, Janicki JS, Wilson JR, Fishman AP (1981) Amrinone and exercise performance in patients with chronic heart failure. Am J Cardiol 48(1):164–169
8. Dahn MS, Lange P, Lobdell K, Hans B, Jacobs LA, Mitchell RA (1987) Splanchnic and total body oxygen consumption differences in septic and injured patients. Surgery 101(1):69–80
9. Carrico CJ, Meakins JL, Marshall JC, Fry D, Maier RV (1986) Multiple-organ-failure syndrome. Arch Surg 121(2):196–208
10. De Backer D, Creteur J, Noordally O, Smail N, Gulbis B, Vincent JL (1998) Does hepato-splanchnic VO2/DO2 dependency exist in critically ill septic patients? Am J Respir Crit Care Med 157(4 Pt 1):1219–1225
11. Reinelt H, Radermacher P, Kiefer P, Fischer G, Wachter U, Vogt J et al (1999) Impact of exogenous beta-adrenergic receptor stimulation on hepato-splanchnic oxygen kinetics and metabolic activity in septic shock. Crit Care Med 27(2):325–331
12. Reinhart K, Bloos F (2005) The value of venous oximetry. Curr Opin Crit Care 11(3):259–263
13. Krantz T, Warberg J, Secher NH (2005) Venous oxygen saturation during normovolaemic haemodilution in the pig. Acta Anaesthesiol Scand 49(8):1149–1156
14. Vincent JL (1992) Does central venous oxygen saturation accurately reflect mixed venous oxygen saturation? Nothing is simple, unfortunately. Intensive Care Med 18(7):386–387
15. Di Filippo A, Gonnelli C, Perretta L, Zagli G, Spina R, Chiostri M et al (2009) Low central venous saturation predicts poor outcome in patients with brain injury after major trauma: a prospective observational study. Scand J Trauma Resusc Emerg Med 17:23
16. Reinhart K, Kuhn HJ, Hartog C, Bredle DL (2004) Continuous central venous and pulmonary artery oxygen saturation monitoring in the critically ill. Intensive Care Med 30(8):1572–1578
17. Herrera A, Pajuelo A, Morano MJ, Ureta MP, Gutierrez-Garcia J, de las Mulas M (1993) Comparison of oxygen saturations in mixed venous and central blood during thoracic anesthesia with selective single-lung ventilation. Rev Esp Anestesiol Reanim 40(6):349–353
18. Ladakis C, Myrianthefs P, Karabinis A, Karatzas G, Dosios T, Fildissis G et al (2001) Central venous and mixed venous oxygen saturation in critically ill patients. Respiration; Inter Rev Thorac Dis 68(3):279–285. Comparative Study
19. Dueck MH, Klimek M, Appenrodt S, Weigand C, Boerner U (2005) Trends but not individual values of central venous oxygen saturation agree with mixed venous oxygen saturation during varying hemodynamic conditions. Anesthesiology 103(2):249–257
20. Pieri M, Brandi LS, Bertolini R, Calafa M, Giunta F (1995) Comparison of bench central and mixed pulmonary venous oxygen saturation in critically ill postsurgical patients. Minerva Anestesiol 61 (7–8):285–291. Comparative Study

21. Rivers E, Nguyen B, Havstad S, Ressler J, Muzzin A, Knoblich B et al (2001) Early goal-directed therapy in the treatment of severe sepsis and septic shock. N Engl J Med 345(19):1368–1377. Clinical Trial Randomized Controlled Trial Research Support, Non-U.S. Gov't

22. Gao F, Melody T, Daniels DF, Giles S, Fox S (2005) The impact of compliance with 6-hour and 24-hour sepsis bundles on hospital mortality in patients with severe sepsis: a prospective observational study. Crit Care 9(6):R764–R770. Comparative Study Evaluation Studies Research Support, Non-U.S. Gov't

23. Pearse R, Dawson D, Fawcett J, Rhodes A, Grounds RM, Bennett ED (2005) Changes in central venous saturation after major surgery, and association with outcome. Crit Care 9(6):R694–R699

24. Rivers E (2006) Mixed vs central venous oxygen saturation may be not numerically equal, but both are still clinically useful. Chest 129(3):507–508

25. Sakr Y, Dubois MJ, De Backer D, Creteur J, Vincent JL (2004) Persistent microcirculatory alterations are associated with organ failure and death in patients with septic shock. Crit Care Med 32(9):1825–1831

26. Legrand M, Bezemer R, Kandil A, Demirci C, Payen D, Ince C (2011) The role of renal hypoperfusion in development of renal microcirculatory dysfunction in endotoxemic rats. Intensive Care Med 37(9): 1534–1542

27. Pope JV, Jones AE, Gaieski DF, Arnold RC, Trzeciak S, Shapiro NI (2010) Multicenter study of central venous oxygen saturation (ScvO(2)) as a predictor of mortality in patients with sepsis. Ann Emerg Med 55(1):40–46. e1

28. Bracht H, Hanggi M, Jeker B, Wegmuller N, Porta F, Tuller D et al (2007) Incidence of low central venous oxygen saturation during unplanned admissions in a multidisciplinary intensive care unit: an observational study. Crit Care 11(1):R2

29. Dellinger RP, Levy MM, Rhodes A, Annane D, Gerlach H, Opal SM et al (2013) Surviving Sepsis Campaign: international guidelines for management of severe sepsis and septic shock, 2012. Intensive Care Med 39(2):165–228. Practice Guideline

30. Mouncey PR, Osborn TM, Power GS, Harrison DA, Sadique MZ, Grieve RD et al (2015) Trial of early, goal-directed resuscitation for septic shock. N Engl J Med 372(14):1301–1311

31. Peake SL, Delaney A, Bailey M, Bellomo R, Cameron PA, Cooper DJ et al (2014) Goal-directed resuscitation for patients with early septic shock. N Engl J Med 371(16):1496–1506

# Echocardiography

5

Echocardiography is one of the monitoring techniques available at the bedside for monitoring the cardiovascular system. Because it is completely noninvasive for transthoracic echocardiography and semi-invasive for transesophageal echocardiography, this technique provides the clinician information on both the anatomical and the functional cardiovascular system. However, this technique remains operator dependent and requires extensive training to correctly perform it; in addition, it has been used only by cardiologists for a long time. The use of echocardiography as a monitoring tool also has its limitations. Indeed, the technique is an evaluation at one time, and this requires the repetition of difficult tests in the clinical setting where the clinician is not always available or not always competent. Thus, echocardiography is most often used as a diagnostic tool or to judge the effect of certain drugs (inotropes, fluid expansion) and never used to monitor during a long time.

The practical use of echocardiography in the ICU is quite different compared with its use in the cardiology community, though the technique is the same [1]. In the ICU, echocardiography is more focused on monitoring and diagnosing a circulatory failure to estimate the cardiac output and ventricular preload. Echocardiography significantly contributes to the anatomical and functional study of the heart and great vessels (aorta, vena cava). The prevailing pressure gradients around the area where it measures the flow velocity are provided by Doppler velocimetry. Doppler velocimetry may be used to estimate the pressure in the pulmonary artery and into the left atrium. The measurement of cardiac output is easily achievable by echocardiography. The evolution of the circulatory condition over time or the response to therapeutic intervention can be evaluated by performing repeated measurements. Its ability to provide a quick etiological diagnosis of shock is one of the greatest advantages of using this technique in the ICU.

## 5.1 Cardiac Output Measurement

Flow measurement is important in certain therapeutic interventions such as volume expansion and inotrope or vasopressor administration. The change in cardiac output in response to therapeutic intervention is a key component of the therapeutic process. The monitoring of changes in the cardiac output requires a monitoring tool [2]; monitoring can easily be achieved with echocardiography and Doppler.

## 5.2 Stroke Volume Measurement

Stroke volume measurement is the most used and validated technique [3]. This measurement is performed by transthoracic echocardiography. The goal is to measure the velocity of

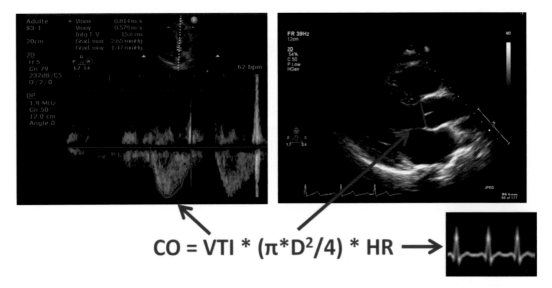

$$CO = VTI * (\pi*D^2/4) * HR \longrightarrow$$

**Fig. 5.1** Cardiac output measurement by transthoracic echocardiography with pulsed Doppler on the apical five-chamber view of the velocity time integral (*VTI*), the diameter (*D*) of the left ventricular outflow tract (LVOT) on the long-axis parasternal view, which is capable of calculating the LVOT surface, and the heart rate (HR) measured on the ECG recording

blood by pulsed Doppler through the aortic valve or directly below the valve at the outflow tract of the left ventricle (Fig. 5.1). The velocity time integral (VTI) is calculated by measuring the envelope of the maximum speed at each instant, which corresponds to the distance traveled by red blood cells during systole eight stroke distance). Then, the multiplication of the VTI by the area of the outflow tract or the valvular orifice provides the stroke volume (Fig. 5.2). The validity of this measurement is achieved only if there is no aortic stenosis or underlying obstacle such as a septal bulge. Because the surface of the outflow tract is fixed, the change in the VTI after a therapeutic intervention allows for an assessment of the change in the stroke volume. By the same principle, it is possible to estimate the cardiac output of the right ventricle by measuring the right ventricle diameter outflow tract and the VTI under the pulmonary valve. However, this method is more complex to perform and is less validated than on the left chambers.

## 5.3    Calculation of the Stroke Volume by Two-Dimensional Echocardiography

The volume of the left ventricular cavity can be measured using simple geometric models. The volume ejected during systole and the ventricular ejection fraction can be calculated by performing these steps in diastole and systole. Various formulae exist. From measurement of the left ventricular diameter, the Teicholz formula estimates the left ventricular volume (Fig. 5.3).

However, Simpson's simplified method, which uses a technique of the successive summation of disks measured at the level of the left ventricular cavity, is the most reliable and most commonly used method. The technique first identifies the contour of the left ventricular cavity and then, according to a predefined algorithm, the long axis of the cavity is determined, and the ventricular cavity is divided into 20 disks over the entire length of the long axis (Fig. 5.4). The ventricular volume is estimated

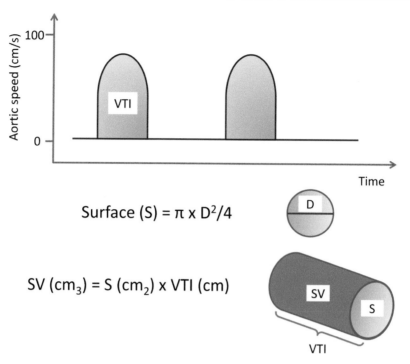

**Fig. 5.2** The principle of calculating the stroke volume by Doppler echocardiography. The subaortic velocity time integral, which corresponds to the amount of blood passing through the LVOT, is provided by the VTI of the flow, which is obtained by tracing the signal envelope. It corresponds to the distance traveled by the fastest red blood cells that cross the LVOT (stroke distance). The diameter of the outflow tract is used to calculate the cross-sectional area, assuming a circular cross section. The product of integrating the time speed by the cross-sectional area corresponds to the stroke volume (volume of a cylinder). The product of stroke volume by the heart rate permits the calculation of the cardiac output

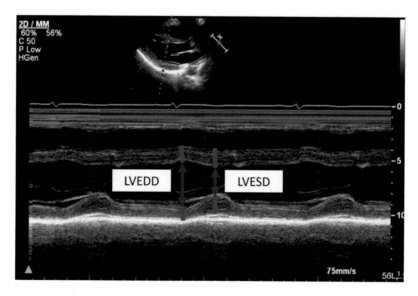

**Fig. 5.3**   The measurement by transthoracic echocardiography (long-axis parasternal view) of left ventricular diameters in time-motion mode. *LVEDD* Left ventricular end-diastolic diameter, *LVESD* left ventricular end-systolic diameter

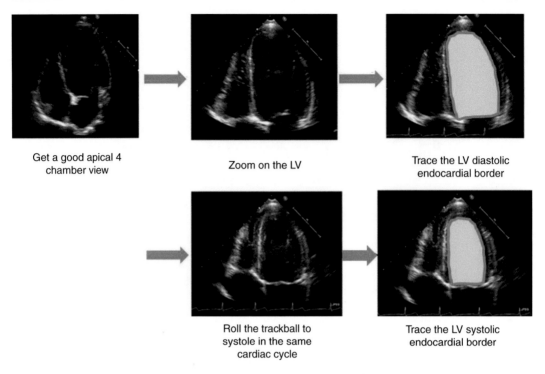

Get a good apical 4 chamber view

Zoom on the LV

Trace the LV diastolic endocardial border

Roll the trackball to systole in the same cardiac cycle

Trace the LV systolic endocardial border

**Fig. 5.4** The principle of measuring the stroke volume and left ventricular ejection fraction by Simpson's method, based on measurement of the volumes of the left ventricular cavity in diastole and systole. Think to perform two perpendicular planes

by adding the volume of each of these disks. The method is more accurate when the measurements are performed in two perpendicular planes. However, it often underestimates the volumes when compared with reference values measured by angiography, which is related to the difficulty in correctly identifying the contour of the endocardium. As the Teicholz formula, this method is also much less accurate when there are disturbances in the left ventricular wall motion.

The most reliable measurement of the left ventricular ejection volume remains Doppler measurement of the aortic blood velocity in association with the measurement of the diameter of the left ventricular outflow tract. This method is the gold standard for estimating the stroke volume in echocardiography. Following a therapeutic intervention (volume expansion, inotropic administration) and to test its efficacy, it is possible to measure only the change in VTI because

the surface of the chamber remains constant. This simple measurement estimates the changes in stroke volume in this context.

## 5.4 Estimation of Pressure Gradients from Doppler

### 5.4.1 Simplified Bernoulli Equation

The principle of energy conservation, with some approximations (i.e., losses, negligible friction, and acceleration phenomena), describes the following relationship between the Doppler speed measurement and the pressure gradient prevailing on either side of the orifice where the measurement of the speed is made. This is the simplified Bernoulli equation, where $P_1$ and $P_2$ are the pressures upstream and downstream of the orifice, respectively, and $V_1$ and $V_2$ are the Doppler speeds upstream and downstream of the orifice, respectively [4].

**Fig. 5.5** Maximum speed measurement of the flow of tricuspid regurgitation for estimating the pressure gradient between the right ventricle and the right atrium according to the simplified Bernoulli equation (transthoracic echocardiography)

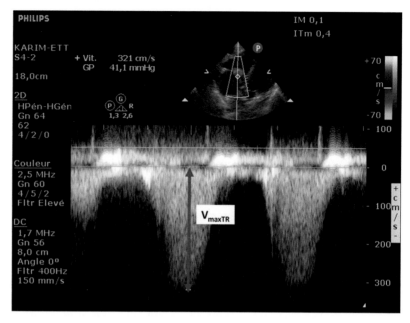

### 5.4.2 Estimated Systolic Pulmonary Artery Pressure

The simplified Bernoulli law (i.e., the Law of Energy Conservation) explains the relationship between the Doppler measurement speed and the pressure gradient between two cavities on either side of an opening:

$$P_1 - P_2 = 4\left(V_2^2 - V_1^2\right)$$

Through the tricuspid valve where physiological regurgitation occurs in systole, it is possible to estimate the pressure through the valve opening. Application of the simplified Bernoulli law to tricuspid regurgitation estimates the pressure that exists on both sides in systole. By applying the above principle to regurgitation through the tricuspid orifice, the pressure can be estimated on either side of this orifice in systole using the formula:

$$\mathrm{RVP} - \mathrm{RAP} = 4\left(V_{\max \mathrm{TR}}\right)^2$$

where RVP is the right ventricular pressure, RAP is the right atrial pressure, and $V_{\max}$ TR is the maximum speed of tricuspid regurgitation. The right

ventricular pressure in systole is very close to the systolic pulmonary artery pressure ($\mathrm{PAP}_{\mathrm{syst}}$) when the pulmonary valve is open (in the absence of pulmonary stenosis). The simplified equation (neglecting the power term because the blood velocity in the right ventricle is small compared with the regurgitation flow velocity) becomes (Fig. 5.5)

$$\mathrm{PAP}_{\mathrm{syst}} = 4\left(V_{\max}\mathrm{TR}\right)^2 + \mathrm{RAP}.$$

## 5.5 Estimating the Filling Pressures of the Left Ventricle

Estimation of the left ventricular filling pressures is important in the case of diastolic heart failure [5] and to differentiate cardiogenic pulmonary edema from an inflammatory pulmonary edema. Echocardiography can identify the existence of high pressure in the left atrium suggesting a cardiogenic pulmonary edema origin. Pulsed Doppler can measure the blood flow velocity, which is proportional to the existing pressure gradient on either side of the Doppler window. When measured at the point of the valve leaflets, the velocity of the mitral filling flow reflects the pressure gradient between the left atrium and ventricle [6]. When the patient is in sinus rhythm, the speed at which the blood moves during this initial action is called the "E-wave" for the early filling velocity. However, some blood always remains; thus, toward the end of the atrial emptying cycle (diastole), the second step occurs in which the atria contract to squeeze out the residual blood ("atrial kick"). The speed of blood filling the ventricle in this step is the "A-wave" for atrial filling velocity. Physiologically, early filling provides two-thirds of the left ventricular filling. The ratio of the peak velocity of the $E$ wave to that of the $A$-wave ($E/A$) is approximately slightly higher than one (Fig. 5.6a). When increasing the pressure of the left atrium, the $E$ wave velocity increases as the pressure gradient between the atrium and the left ventricle increases. In addition, the deceleration time of the $E$ wave (DT; the time between the peak and return point of the $E$ wave at baseline) is shortened (less than 120 ms) because of rapid pressure equalization between the atrium and the left ventricle. In this case, the part of filling that is provided by atrial contraction decreases the $E/A$ ratio [6]. The TDE increases in hypovolemia and decreases in hypervolemia or in the case of disorder in left ventricle relaxation.

It is also possible to estimate the pulmonary venous flow by placing the pulsed Doppler window at the junction between the pulmonary vein and the left atrium. A three-phase flow with a small regurgitation wave during atrial contraction is then displayed. There are then two anterograde systolic and diastolic waves. The systolic component is normally predominant in healthy subjects.

Another Doppler technique is the use of tissue Doppler (TD) to assess the diastolic velocity movement of the mitral annulus. This component is a reflection of active and longitudinal LV relaxation. After adjusting the ultrasound on tissue Doppler (TD) with a low-frequency transmission (≤4 MHz), the pulsed Doppler window is adjusted to the level of the mitral annulus with an opening of 5–10 mm, and scale velocities are set between 15 and 20 cm/s. The Doppler axis is aligned as much as possible relative to movement of the mitral annulus. Slight changes are sometimes required. The normal $E$ velocity is approximately 10–15 cm/s in young adults and 8 cm/s in the elderly (>70 years) and varies depending on where the measurement is conducted, as follows: >8 cm/s at the septal level and >10 cm/s at the lateral side level. Because the conventional pulsed Doppler "E-wave" for the early filling velocity is dependent on both volemia and myocardial proprieties during diastole (relaxation) and because TD at the mitral annulus $E'$ measures only the myocardial proprieties during diastole, the ratio $E/E'$ is used to evaluate the left atrial preload (volemia). As the ratio increases, the LAP also increases [7]. This relationship no longer exists when the mitral ring calcifies or the mitral valve is diseased [8]. The major limitation of this measure is that it tends to define diastolic function of the entire ventricle from a measure that concerns only the longitudinal early diastolic relaxation of the side wall, septal or anterior. Moreover, this measure does not evaluate the LV passive compliance [9, 10] (Fig. 5.6b).

• Planimetry of the left atrium allows assessment of the LA volume. Its elevation to more than 37 ml/m$^2$ corresponds to a sustained elevation of the LAP in relation to LV diastolic failure [11]. It is also a prognostic marker. The risks of death, heart failure, atrial fibrillation, and ischemic stroke significantly increase when the LA volume exceeds this value [12].

## 5.6    Assessment of Right Ventricular Function

Because the right ventricular preload and afterload vary with respiration, measurements concerning the right ventricle must be made at the end of expiration or during apnea. Unlike the left ventricle, the right ventricle has a more complex form and cannot be described with a simple geometric figure. Therefore, it is impossible to calculate the ejection fraction based on volumetric methods. In addition, the contour of the endocardium is very difficult to model because of the many trabeculae present in its cavity. Finally, the data provided in the literature were obtained by transthoracic

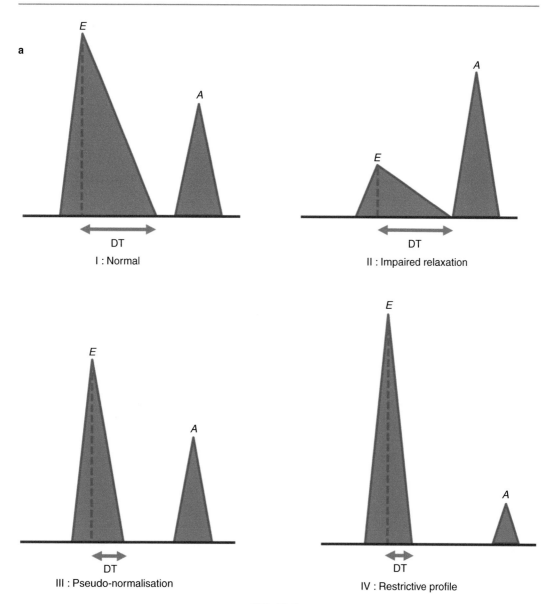

**Mitral inflow**

**Fig. 5.6** Changes in the mitral inflow (**a**) and the mitral annulus velocity (**b**) according to healthy patients and patients with diastolic dysfunction. Stage II corresponds to a relaxation defect, decreased mitral flow velocity liabilities (*E*), and increased atrial contribution. However, a decrease in the mitral annulus velocity occurs. Stage III corresponds to a pseudo-normalization reflecting the gradual increase in left atrial pressure that restores the LA-LV pressure gradient. As the flow velocity increases, *E* returns to its normal value. However, its morphology is pathological, with the acceleration and deceleration slopes accentuated (DT <150 ms) and the isovolumetric relaxation shortened (IVRT <200 ms). The mitral annulus velocity decreases. Stage IV corresponds to a "restrictive" profile: this is the most serious situation. The ventricle becomes stiff, and the atrial pressure increases as the flow velocity *E* becomes very large with a very fast deceleration (DT <100 ms). The atrial contraction wave becomes almost negligible because the ventricle is not distensible at the end of diastole. The velocity of the mitral annulus is very limited

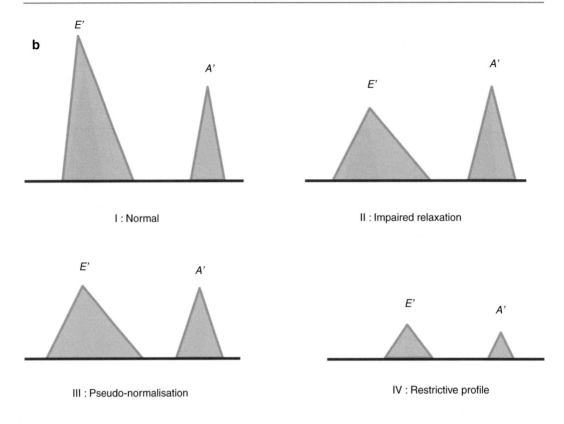

**Mitral annulus velocity**

**Fig. 5.6** (continued)

echocardiography during spontaneous breathing but were not validated for TEE in patients ventilated with positive pressure ventilation.

Systolic RV function is a major determinant of the clinical outcome of critically ill patients. It is therefore essential to be able to measure its whole function. The right ventricular ejection is performed in a low flow resistance system. Therefore, the RV is very sensitive to sudden increases in afterload with a consequent expansion of the cavity and a loss of systolic function. Thus, during acute PAH (pulmonary embolism, hypoxic pulmonary vasoconstriction, or ARDS), the RV abruptly expands. In an inextensible pericardial sac, the biventricular volume is constant. Therefore, RV dilation under acute cor pulmonale (ACP) results in left ventricular compression and a "paradoxical septum," i.e., bulging of the interventricular septum into the left ventricle. The paradoxical septum partially

occludes the left ventricle, causing a decrease in cardiac output on the left side and, consequently, a drop in systemic blood pressure followed by hemodynamic instability. An ultrasound diagnosis of ACP is based on the finding that RV dilation is associated with a paradoxical septum that compresses the LV (interventricular interactions) [13]. A ratio between the diastolic ventricular surfaces (RV/LV) of >0.6 suggests RV dilation [14] (Fig. 5.7).

The right ventricular outflow tract (RVOT) shortening fraction is calculated by the difference between the diastolic and systolic diameters divided by the end-diastolic diameter. There is a wide interindividual variation making its use rather difficult in clinical practice. The right ventricular outflow tract (RVOT) ejection fraction is calculated as the difference between the end-diastolic volume and the end-systolic area, divided by the end-diastolic volume. However, as

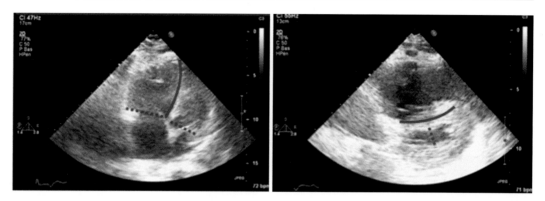

**Fig. 5.7** Ultrasound example (transthoracic echocardiography sectional parasternal long axis (*left*) and short axis (*right*) in diastole) of the right heart failure in the context of acute cor pulmonale after massive pulmonary embolism with severe RV dilatation, an RV/LV ratio >0.6, a paradoxical septum (*solid line*), and LV compression

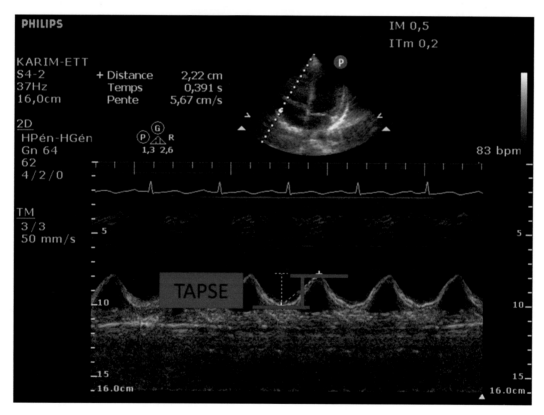

**Fig. 5.8** Tricuspid annular plane systolic excursion (*TAPSE*) measurement by transthoracic echocardiography in apical four-chamber mode and then in time-motion mode corresponding to the systolic excursion of the lateral tricuspid ring, measured in cm

the fraction of the diameter decreases, the shortening fraction of the surface is not very useful in the ICU. Shortening fraction of a diameter or a surface of <0.3 indicates severe right ventricular failure and is associated with worsening perioperative mortality [15]. All these markers of right ventricular function have been validated against the FE measured by MRI [16–18].

The tricuspid annular plane systolic excursion (TAPSE) is the longitudinal displacement of the side portion of the tricuspid annulus [19] (Fig. 5.8). Because the longitudinal shortening

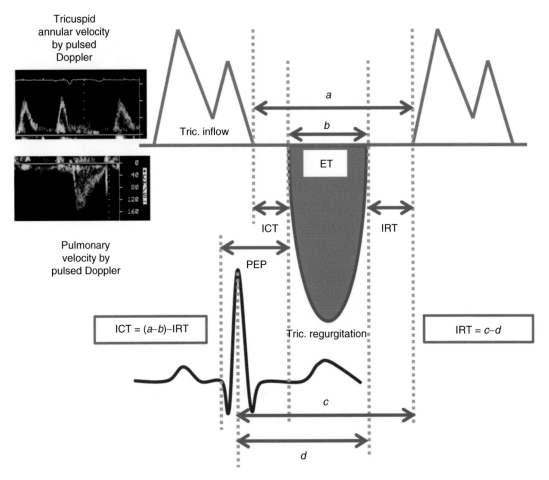

**Fig. 5.9** Schematic representation of the Doppler measurement intervals for calculation of the Tei index, the myocardial performance index of the RV. It is defined as the sum of the isovolumetric contraction time (*ICT*), the ejection time (*ET*), and the isovolumetric relaxation (*IRT*) divided by the ejection time $(a-b)/b$. The normal value is 0.5 [26]. The Tei index is calculated on the Doppler spectral display of tricuspid and pulmonary flow

is the major component of the right ventricular contraction, the present fact makes this measure relevant in evaluating the contractile function of the right ventricle. A value of greater than 15 mm indicates good RV systolic function [20]. However, this measure has two weaknesses: it may overestimate the real RV function (because the apex can be driven by the septum and the LV contraction). Differently, the systolic velocity of the tricuspid annulus assessed with tissue Doppler allows measurements to estimate the RV systolic function. This measure is validated for TEE [21] and for

patients under positive pressure ventilation [22]. A maximum speed of less than 7 cm/s is associated with an EF <0.4 [7]. However, the maximum speed is lower when the measurement is performed by TEE than by transthoracic echocardiography [23].

The acceleration of isovolumetric contraction measured at the tricuspid annulus is currently the most reliable index because of the few dependencies on load conditions. However, tissue Doppler requires a high sampling frequency (i.e., strain mode rate) [16, 17, 24]. The highest tissue velocities (VDs) are faster at the base

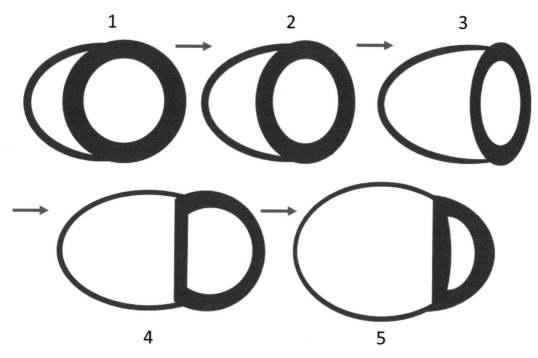

**Fig. 5.10** Schematic representation of the occurrence of right ventricular failure and its impact on the left ventricle, the interventricular septum, and RV/LV interdependence in five steps: *1* normal biventricular configuration; *2* the start of right ventricular dilation and wall thinning; *3* RV dilation, further thinning of the RV walls, flattening of the interventricular septum, and reduction in the LV volume; *4* significant dilation and thinning of the RV free wall and correctness of the interventricular septum; and *5* RV failure, severe RV dilation cavity, paradoxical septal bulging into the LV cavity, and severe reduction in LV volume [17]

than at the apex and are significantly higher than those of the LV.

Also, the myocardial performance Tei index is applicable to the RV function assessment (Fig. 5.9). It is the ratio between the duration of the isovolumetric phases and the duration of ejection [17]. A severe dysfunction is characterized by an index >0.5. It is also possible to calculate the Tei index from tissue Doppler on the tricuspid annulus. Although it represents only longitudinal RV contraction, this measurement can be made during a single cardiac cycle.

The acceleration slope of tricuspid regurgitation (TR) is the pressure difference as a function of time of the isovolumetric contraction phase of the RV; this measurement which is equivalent to a d$p$/d$t$ of the right ventricle has been validated to assess the RV systolic function [25].

In clinical practice, right ventricular echocardiographic measurements may be inadequate because of an incorrect interpretation of ventricular contours. Therefore, all indices must be integrated as part of an overall assessment of ventricular function and the clinical situation. A single result should not in itself lead to therapeutic decisions. In the case of right ventricular failure, the right ventricle expands due to the thinness of its walls (Fig. 5.10). Therefore, a small RV de facto excludes a right ventricular failure, with the exception of RV myocardial infarction and in case of large pericardial effusion. The measurements of longitudinal shortening by the TAPSE and of free wall thickening are the two essential measures to take into account in assessing the right ventricle. Finally, notably, the gold standard for evaluating the right ventricle is the MRI [27, 28].

# References

1. Beaulieu Y (2007) Bedside echocardiography in the assessment of the critically ill. Crit Care Med 35(5 Suppl):S235–S249. [Review]

2. Giraud R, Siegenthaler N, Tagan D, Bendjelid K (2011) Evaluation of skills required to practice advanced echocardiography in intensive care. Rev Med Suisse 7(282):413–416

3. Huntsman LL, Stewart DK, Barnes SR, Franklin SB, Colocousis JS, Hessel EA (1983) Noninvasive doppler determination of cardiac output in man. Clinical validation. Circulation 67(3):593–602

4. Lang RM, Tsang W, Weinert L, Mor-Avi V, Chandra S (2011) Valvular heart disease. The value of 3-dimensional echocardiography. J Am Coll Cardiol 58(19):1933–1944

5. Aurigemma GP, Gaasch WH (2004) Clinical practice. Diastolic heart failure. N Engl J Med 351(11): 1097–1105

6. Appleton CP, Hatle LK, Popp RL (1988) Relation of transmitral flow velocity patterns to left ventricular diastolic function: new insights from a combined hemodynamic and Doppler echocardiographic study. J Am Coll Cardiol 12(2):426–440

7. Skubas N (2009) Intraoperative Doppler tissue imaging is a valuable addition to cardiac anesthesiologists' armamentarium: a core review. Anesth Analg 108(1): 48–66

8. Nagueh SF, Appleton CP, Gillebert TC, Marino PN, Oh JK, Smiseth OA et al (2009) Recommendations for the evaluation of left ventricular diastolic function by echocardiography. Eur J Echocardiogr 10(2):165–193

9. Nagueh SF, Middleton KJ, Kopelen HA, Zoghbi WA, Quinones MA (1997) Doppler tissue imaging: a non-invasive technique for evaluation of left ventricular relaxation and estimation of filling pressures. J Am Coll Cardiol 30(6):1527–1533

10. Ommen SR, Nishimura RA, Appleton CP, Miller FA, Oh JK, Redfield MM et al (2000) Clinical utility of Doppler echocardiography and tissue Doppler imaging in the estimation of left ventricular filling pressures: a comparative simultaneous Doppler-catheterization study. Circulation 102(15):1788–1794

11. Pritchett AM, Mahoney DW, Jacobsen SJ, Rodeheffer RJ, Karon BL, Redfield MM (2005) Diastolic dysfunction and left atrial volume: a population-based study. J Am Coll Cardiol 45(1):87–92

12. Abhayaratna WP, Seward JB, Appleton CP, Douglas PS, Oh JK, Tajik AJ et al (2006) Left atrial size: physiologic determinants and clinical applications. J Am Coll Cardiol 47(12):2357–2363

13. Vieillard-Baron A, Prin S, Chergui K, Dubourg O, Jardin F (2002) Echo-Doppler demonstration of acute cor pulmonale at the bedside in the medical intensive care unit. Am J Respir Crit Care Med 166(10):1310–1319

14. Jardin F, Dubourg O, Bourdarias JP (1997) Echocardiographic pattern of acute cor pulmonale. Chest 111(1):209–217

15. Maslow AD, Regan MM, Panzica P, Heindel S, Mashikian J, Comunale ME (2002) Precardiopulmonary bypass right ventricular function is associated with poor outcome after coronary artery bypass grafting in patients with severe left ventricular systolic dysfunction. Anesth Analg 95(6):1507–1518, table of contents

16. Haddad F, Couture P, Tousignant C, Denault AY (2009) The right ventricle in cardiac surgery, a perioperative perspective: I. Anatomy, physiology, and assessment. Anesth Analg 108(2):407–421

17. Haddad F, Doyle R, Murphy DJ, Hunt SA (2008) Right ventricular function in cardiovascular disease, part II: pathophysiology, clinical importance, and management of right ventricular failure. Circulation 117(13):1717–1731

18. Lang RM, Bierig M, Devereux RB, Flachskampf FA, Foster E, Pellikka PA et al (2006) Recommendations for chamber quantification. Eur J Echocardiogr 7(2):79–108

19. Hammarstrom E, Wranne B, Pinto FJ, Puryear J, Popp RL (1991) Tricuspid annular motion. J Am Soc Echocardiogr 4(2):131–139

20. Kaul S, Tei C, Hopkins JM, Shah PM (1984) Assessment of right ventricular function using two-dimensional echocardiography. Am Heart J 107(3):526–531

21. David JS, Tousignant CP, Bowry R (2006) Tricuspid annular velocity in patients undergoing cardiac operation using transesophageal echocardiography. J Am Soc Echocardiogr 19(3):329–334

22. Michaux I, Filipovic M, Skarvan K, Schneiter S, Schumann R, Zerkowski HR et al (2006) Effects of on-pump versus off-pump coronary artery bypass graft surgery on right ventricular function. J Thorac Cardiovasc Surg 131(6):1281–1288

23. Tunick PA, Kronzon I (2000) Atheromas of the thoracic aorta: clinical and therapeutic update. J Am Coll Cardiol 35(3):545–554

24. Vogel M, Schmidt MR, Kristiansen SB, Cheung M, White PA, Sorensen K et al (2002) Validation of myocardial acceleration during isovolumic contraction as a novel noninvasive index of right ventricular contractility: comparison with ventricular pressure-volume relations in an animal model. Circulation 105(14):1693–1699

25. Anconina J, Danchin N, Selton-Suty C, Isaaz K, Juilliere Y, Buffet P et al (1992) Measurement of right ventricular dP/dt. A simultaneous/comparative hemodynamic and Doppler echocardiographic study. Arch Mal Coeur Vaiss 85(9):1317–1321

26. Tei C, Nishimura RA, Seward JB, Tajik AJ (1997) Noninvasive Doppler-derived myocardial performance index: correlation with simultaneous measurements of cardiac catheterization measurements. J Am Soc Echocardiogr 10(2):169–178

27. Keenan NG, Pennell DJ (2007) CMR of ventricular function. Echocardiography 24(2):185–193

28. Maceira AM, Prasad SK, Khan M, Pennell DJ (2006) Reference right ventricular systolic and diastolic function normalized to age, gender and body surface area from steady-state free precession cardiovascular magnetic resonance. Eur Heart J 27(23):2879–2888

# Preload Dependency Dynamic Indices

Volume expansion was formerly assessed by invasive cardiac output monitoring and intracardiac pressure measurement, following the concept of "fluid challenge" [1]. Recently, fluid responsiveness was then evaluated in response to volume expansion. The recent concept is based on the Starling curve. For instance, following a volume expansion, a patient may be situated on the ascending portion of the curve (with a significant increase in cardiac output without massive increases in filling pressures) or may be located on the flat portion of the curve (with a small increase in cardiac output along with drastically increased filling pressures) [2]. In the past, no indication was predictive of the potential effectiveness of volume expansion, and the only evidence used to guide fluid therapy was the preload indices, also known as "static" indices. However, the central venous pressure (CVP) and pulmonary artery occlusion pressure (PAOP) were never shown to reliably predict the hemodynamic benefit of volume expansion [3, 4] to a greater extent than ventricular diameter or surface measurements [5, 6]. Although a low preload value could still achieve a volume expansion in shocked (and not yet resuscitated) patients, the use of static preload indices is currently not recommended to guide fluid therapy in the ICU. In fact, this practice is likely to lead to the incorrect administration of intravenous fluids and to result in an increased risk or aggravation of pulmonary edema, hypoxemia, or ARDS [7, 8].

A new approach based on hemodynamic cardiopulmonary interactions has been developed in the past decade [8]. Mechanical ventilation causes lung volumes to phasically vary by applying positive airway pressure during inspiration above the resting elastic recoil pressure of the lung and chest wall. The resulting phasic increases in lung volume cause proportional increases in intrathoracic pressure (ITP) as the expanding lungs press against the chest wall and diaphragm. The net result of these cyclic changes in ITP is parallel changes in the right atrial pressure. Because the right atrial pressure is the downstream pressure for venous return, increasing the right atrial pressure transiently decreases this pressure gradient; accordingly, venous return to the right ventricle decreases, and intrathoracic blood volume decreases during lung inflation [9]. After approximately three heart beats, the decreased flow reaches the left ventricle; if it is preload responsive, then its output also transiently decreases but that occurs during expiration [10]. Thus, observing either the left ventricular stroke volume or its surrogate, arterial pulse pressure variation (SVV and PPV, respectively), during ventilation identifies patients as volume responsive. This fundamental observation, made over 30 years ago, has been rediscovered as a form of functional hemodynamic monitoring [11] and used to determine whether patients in shock will increase their cardiac outputs if given a fluid challenge [12].

© Springer International Publishing Switzerland 2016
R. Giraud, K. Bendjelid, *Hemodynamic Monitoring in the ICU*, DOI 10.1007/978-3-319-29430-8_6

The principle is to evaluate whether volume expansion could be beneficial to the patient prior to any corrective action. There are then two categories of patients: "responders," in whom volume expansion, if practiced, will significantly increase the cardiac output, and "nonresponders," in whom no beneficial significant hemodynamic improvement can be expected with volume infusion. The concept is to use dynamically "disruptive" heart-lung interactions to test the evolution of hemodynamic parameters such as mechanical ventilation or passive leg raising.

## 6.1 Passive Leg Raising to Test the Preload Dependency

An alternative for assessing volume expansion needs is the passive leg raising maneuver (Fig. 6.1).

This is particularly the case when the respiratory variation in the hemodynamic indices loses its predictive ability to test the response to fluid loading. These situations are identified as follows: low tidal volumes, cardiac arrhythmias, and spontaneous breathing. This maneuver comprises simultaneously lifting both of the patient's lower limbs to an angle of 45° to allow for the transfer of venous blood contained in the legs to the intrathoracic compartment [14]. This causes increased right ventricular [15] and then left ventricular preloads [16, 17], and the result of this maneuver is similar to an autotransfusion, mimicking the effects of fluid expansion. This technique is well known among anesthetists and emergency doctors because it is an emergency maneuver used for patients in hypovolemic shock. The passive leg raising increases the cardiac preload by increasing the average circulatory pressure and increases the driving pressure of the venous return [18] by the gravitational movement of blood. This venous blood then passes from the unstressed venous compartment to the stressed venous compartment.

The amplification of this phenomenon is observed in patients under positive pressure mechanical ventilation. Indeed, the volume of blood present in the thoracic and splanchnic venous beds is already constrained by the positive pressure applied by the ventilator. Hence, there is a greater effect in the passive leg raising maneuver for ventilated patients than in spontaneously breathing patients [19]. In contrast, the effects of this maneuver on cardiac output are variable [20] and are dependent on the degree of the elevation of the lower limbs and the volume of the preload reserve. Indeed, some authors have tested the effect of passive leg raising on the increase in stroke volume in patients under mechanical ventilation [21]. Patients with an increased stroke volume during this maneuver had an increase in stroke volume after a fluid loading of 300 ml of IV saline. In contrast, patients in whom the passive leg raising did not lead to increased stroke volumes did not increase their stroke volume after receiving the same volume of fluid loading [21].

Fluid loading

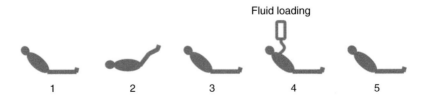

1          2          3          4          5

**Fig. 6.1** Schematic representation of the passive leg raising maneuver. During this maneuver, the cardiac output or one of its derivatives (e.g., ΔPP, SVV, etc.) are continuously monitored. (*1*) The patient is seated at 45° in a semi-supine position. (*2*) The passive leg raising maneuver is conducted by rocking the bed; the patient's chest is flat, and the legs are raised to 45°. Note that the angle between the thoracoabdominal area and lower limbs does not vary during the maneuver. The cardiac output or derivative indices are compared with values before the maneuver. (*3*) The patient is returned to a semi-sitting position. At this stage, if the patient is a "responder," he will receive a fluid infusion (*4*); otherwise, the test stops at step 3. (*5*) The test is ended, and volume expansion is assessed by measuring the cardiac output or one of its derivatives to observe the benefit of fluid expansion (Adapted from Monnet et al. [13])

The passive leg raising maneuver is now used as a test to detect the existence of preload dependency [13]. In the same study, the authors observed the effects of passive leg raising on filling pressures, which return to their original values when the legs are returned to the horizontal position. One of the undeniable advantages of this maneuver is the prevention of deleterious fluid administration. Although initially tested on changes in the stroke volume, the passive leg raising maneuver can also be tested on changes in the pulse pressure. Indeed, pulse pressure variations are directly proportional to the stroke volume, and assuming that arterial compliance is not changed by the maneuver, an increased pulse pressure during passive leg raising would indicate that the stroke volume increased during the maneuver, thus allowing the clinician to conclude that the patient is preload dependent and that fluid expansion may be helpful. This was shown in a study of mechanically ventilated patients, in whom the increase in pulse pressure predicts fluid responsiveness [21]. However, the correlation between changes induced by passive leg raising on the stroke volume and on pulse pressure was not optimal. Indeed, in this study, the blood pressure was measured on the radial artery, and changes in the radial pulse pressure did not optimally reflect the changes in the pulse pressure measured at the aortic level. The changes in stroke volume may, as part of this maneuver, be measured by echocardiography [22], by esophageal Doppler [13], or by pulse contour analysis [23].

## 6.2   Using the Effects of Mechanical Ventilation on Hemodynamic Parameters

The use of positive pressure mechanical ventilation, which leads to cyclical changes in transpulmonary pressure and pleural pressure, causes changes in the hemodynamic parameters of the right and left heart function [2, 24]. With greater respiratory variations, the heart is more significantly "preload dependent," and a volume expansion will affect the patient's hemodynamic status.

During the respiratory cycle under positive pressure, positive pressure leads to an increase of the pleural pressure, which reduces venous return to the right heart and its preload. When the RV is "preload dependent," a decrease in the right ventricular ejection fraction will occur. Meanwhile, the increase in transpulmonary pressure in the insufflation will lead to an increase in RV afterload, leading to a decrease in the right ventricular stroke volume. Therefore, during the mechanical insufflation, by combining these two effects, with a predominant effect of pleural pressure effect on the preload [25, 26], a decline in the RV stroke volume that correlates with the preload dependency of the right heart will occur [2]. By reducing the right ventricular SV, the impact of mechanical breath on the left ventricle will occur in a delayed manner, approximately three to four cardiac cycles later, due to the long transit time of blood in the pulmonary circulation. And the decrease of stroke volume of the right ventricle observed earlier will affect the left ventricular filling after. Therefore, the decrease in the left ventricular preload occurs during the expiratory phase. Accordingly, a patient with a preload dependency will be identified by a respiratory change in the left ventricular ejection volume relative to the inspiratory decrease in RV stroke volume [27].

## 6.2.1   The Respiratory Variation of Systolic Blood Pressure

After performing a respiratory pause to measure the systolic blood pressure as a reference, this method is used to calculate the difference between the minimum and maximum systolic pressures during a respiratory cycle [28]. However, as the respiratory changes in the systolic pressure (SPV) are multifactorial, the method is divided into two distinct parts, using the systolic blood pressure measured during an end-expiratory pause as a reference (Fig. 6.2). The Δup component is equal to the difference between the maximum systolic blood pressure during the respiratory cycle and the reference blood pressure; the Δ down component is equal

**Fig. 6.2** Principles of the measurements of Δup, Δdown, and the change in systolic blood pressure (*SPV*) from the respiratory variations in blood pressure induced by mechanical positive pressure ventilation

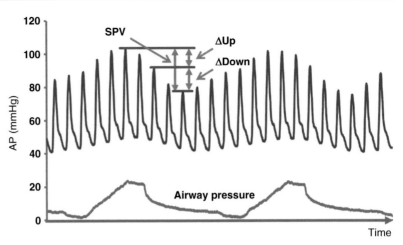

to the difference between the reference systolic blood pressure and the minimum systolic blood pressure during the respiratory cycle. The latter component reflects the expiratory reduction in the left ventricular stroke volume secondary to the inspiratory reduction of venous return and the right ventricular stroke volume, which gives an indication of preload dependency.

Calculation of the SPV is performed according to the following formula [29, 30]:

$$SPV(\%) = \frac{SBP_{max} - SBP_{min}}{(SBP_{max} + SBP_{min})/2} \times 100$$

Δdown is calculated according to the following formula:

$$\Delta Down(\%) = \frac{SBP_{reference} - SBP_{min}}{SBP_{reference}} \times 100$$

According to several studies, these parameters are sensitive indicators of preload dependency [27, 31–34]. This was recently confirmed by clinical studies [6, 35–37], which revealed a relationship among the degree of hypovolemia, the amplitude of the SPV, and the value of Δdown. These indices were also significantly more reliable in predicting fluid responsiveness than the PAOP or the diastolic area of the left ventricle. A threshold value of 4.5 % for Δdown predicts a positive fluid response, with useful predictive values. A threshold value of 10 mmHg can be used for the SPV. Finally, there is a significant

relationship between the Δdown value at baseline and the increased stroke volume secondary to volume expansion with a good quantitative estimation of fluid responsiveness.

### 6.2.2 Measurement of Pulse Pressure Respiratory Variations (ΔPP)

The PP increases during insufflation and decreases during expiration under mechanical ventilation using positive pressure. Because the PP is directly correlated to the left ventricular stroke volume, those variations reflect only respiratory variations of this volume. In the case of hypovolemia, a lower systemic venous return at each insufflation is more pronounced. As a result, an inspiratory decrease of the right ventricular stroke volume and, a few heartbeats later, an expiratory decrease of the pulmonary venous return, the left ventricular stroke volume, and the PP are observed. As the first physiological indicator of cardiorespiratory interactions, the ΔPP has been validated in clinical practice as an index that predicts fluid responsiveness [33] (Fig. 6.3). A ΔPP greater than 13 % reliably predicts a significant increase in the cardiac output produced by a volume expansion of 500 ml [33]. It is calculated using the following formula:

$$\Delta PP(\%) = \frac{PP_{max} - PP_{min}}{(PP_{max} + PP_{min})/2} \times 100$$

**Fig. 6.3** Principle of the measurement of respiratory variation in the arterial pulse pressure (ΔPP) in a patient on mechanical positive pressure ventilation and sinus rhythm

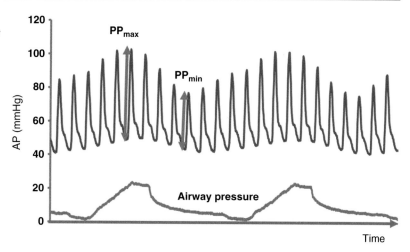

Other studies have confirmed the calculation of ΔPP in predicting fluid responsiveness in various types of patients [6, 30, 31, 34–36, 38]. However, there are a number of limitations in the use of this index. It is mandatory that patients receiving mechanical ventilation do not trigger the ventilator, the heart rhythm must be regular, and the tidal volume must be greater than 8 ml/kg of the predicted body weight [39–42]. It has also been shown that the abdominal pressure [43], the respiration rate [39], and the presence of right ventricular failure [24] can influence the ΔPP value. This greatly limits the use of ΔPP in clinical practice, especially in ARDS patients ventilated using a protective ventilation strategy (i.e., with a tidal volume less than 6 ml/kg of the predicted body weight).

### 6.2.3   Measurement of Respiratory Stroke Volume Variations

Echocardiography measures the left ventricular SV beat by beat by multiplying the velocity time integral (VTI) obtained by Doppler by the surface of the left ventricular outflow tract [44]. The amplitude of the VTI respiratory variations correlates with the degree of hypovolemia and predicts fluid responsiveness [45] (Fig. 6.4). Esophageal Doppler permits blood flow monitoring in the descending aorta and uses a similar index to predict fluid responsiveness [46]. Changes in the left ventricular SV are reflected

by changes in the maximum flow velocity with equivalent results [5]. The threshold values of these indices should not be used alone but rather in conjunction with clinical findings and the overall hemodynamic situation.

One method to assess the left ventricular SV beat by beat is pulse contour analysis of the arterial pressure wave. This method requires an arterial line and a monitoring device containing a specific algorithm. Many systems allow for real-time analysis with automated calculation of the amount of variation of the left ventricular SV in seconds. The SVV is calculated using the following formula (Fig. 6.5):

$$SVV(\%) = \frac{SVL_{max} - SV_{min}}{(SV_{max} + SV_{min})/2} \times 100$$

As is the case for ΔPP, these systems suffer from limitations, including mechanical ventilation without spontaneous breathing movement and sinus rhythm. Nonetheless, new algorithms allow other devices to overcome extrasystole and predict fluid responsiveness despite these artifacts with good precision [47].

### 6.2.4   Pulsed Plethysmography

By analyzing the pulsatility of the plethysmography curve, similar respiratory variations were observed in the pulse pressure; accordingly, variability was

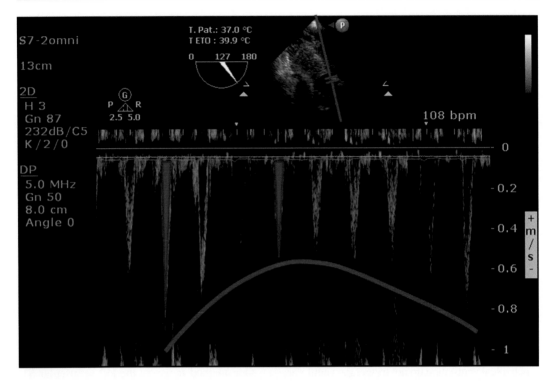

**Fig. 6.4** Transesophageal Doppler echocardiography. Transgastric 127° view of the left ventricular outflow tract showing the respiratory variation of the velocity time integral (VTI)

**Fig. 6.5** The principle of measurement of the respiratory variation in stroke volume (SVV) in a patient on positive pressure mechanical ventilation and sinus rhythm

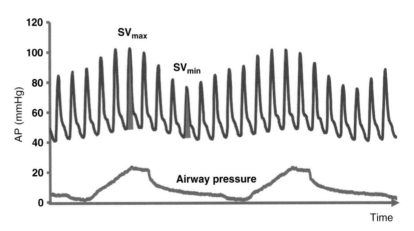

also observed in respiratory pulse plethysmography [30, 48] (Fig. 6.6). Plethysmographic indices include respiratory variation of the amplitudes of the plethysmographic pulse oximetry (ΔPleth) and the plethysmographic variability index (PVI). Both ΔPleth and PVI are obtained through continuous analyses of pulse oximetry signals and calculated using the following formula:

$$\Delta \text{Pleph}(\%) = \frac{\text{Pleph}_{max} - \text{Pleph}_{min}}{\left(\text{Pleph}_{max} + \text{Pleph}_{min}\right)/2} \times 100$$

The pulse pressure signal and "pulse-plethysmography" curves show excellent correlation in several studies. They indirectly and noninvasively assess respiratory variations in the SV, therefore indicating the preload dependency

**Fig. 6.6** The simultaneous recording of systemic
arterial pressure, plethysmography, and respiration
curves of a mechanically
ventilated patient showing
s i g n i f i c a n t
plethysmography respira-
tory variations, indicating
preload dependency

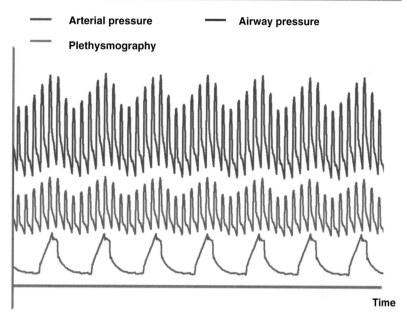

and predicting the fluid responsiveness [48, 49].
Pulse plethysmography can also detect preload
dependency via the passive leg raising maneuver
[50]. However, this method, although noninvasive, suffers from many limitations. As is the case
with other derivatives of SV variations, the present marker should be performed in patients with
mechanical ventilation, in regular rhythm, and
with a sufficient tidal volume due to the lack of a
reliable threshold value in predicting fluid
responsiveness [51].

Indices derived from the arterial pressure
curve are not reliable in patients with spontaneous breathing and who are experiencing an
arrhythmia [52]. In those circumstances, the plethysmographic indices, which share the same
physiological basis of the derivatives of arterial
waveform analysis, lose their predictive ability.
No studies on the plethysmographic indices have
been conducted in arrhythmic patients. However,
two studies [53, 54] show data confirming that
the plethysmographic index most likely has limited value in spontaneously breathing patients; in
these patients, a passive leg raising maneuver
should instead be used to predict fluid responsiveness [55]. As they are based on noninvasive
pulse oximetry, the plethysmographic indices do
not require arterial catheterization and allow for

the safer and faster evaluation of preload
dependency based on indices derived from the
arterial pressure curve. However, the quality of
the plethysmographic signal depends essentially
on peripheral perfusion [56], which can be
significantly reduced by factors such as hypothermia [57], low cardiac output [58], and
vasopressor use.

In particular, norepinephrine use and increasing peripheral vascular tone can reduce the
pulsatile component of the plethysmographic
waveform and thus the precision of the plethysmographic indices. Observational studies [59]
have shown that the correlation between ΔPpleth/
PVI and ΔPP is significantly reduced in patients
receiving norepinephrine. Finally, a study on critically ill patients using laser velocimetry devices
revealed spontaneous cyclic changes in vasomotor tone that can cause periodic oscillations of the
scale of the registered plethysmographic signal
[60]. These oscillations are superimposed on the
plethysmographic respiratory variation signal
and may cause interference, which requires
appropriate signal treatment to be eliminated
[61]. Another factor that could affect the precision of plethysmographic indices is the measurement site. In one study [62], the accuracy of PVI
as a predictor of fluid responsiveness was higher

when the probe was placed on the forehead or the ear lobe where the subcutaneous vascular system is relatively resistant to the sympathetically mediated induction of a change of vasomotor tone [63]. This suggests that in contrast to the usual practice, the finger is not the preferred measurement site for plethysmographic indices, particularly in patients receiving vasoactive drugs.

Signal filtration is also an important technical limitation of plethysmographic indices. Indeed, pulse oximeters have mainly been developed to detect the oxygen saturation signal rather than to detect changes induced by ventilation. In particular, the adaptive digital filters that dynamically change their filter characteristics in response to a change in the sound of the band were adopted to eliminate noise occurring within the bandwidth of the pulse oximetry signal. However, the ongoing adjustment in the filter can lead to changes in the amplitude of the pulse that interfere with those induced by mechanical ventilation [64]. No difference in accuracy between the PVI and the ΔPpleth was determined in a meta-analysis. Similarly, the only study to date that has directly evaluated the PVI/ΔPpleth ratio found no significant difference between these two indices [65].

### 6.2.5   Measurements of Inferior Vena Cava Respiratory Variations (ΔICV)

An ultrasound examination of the IVC can be easily performed by transthoracic echocardiography, particularly using a subcostal view [66]. The IVC diameter measurement discriminate healthy subjects from patients with a high RAP [67]. In his famous study conducted on venous return in dogs, Guyton showed that a moderately negative RAP increased venous return but that greater negativity, in contrast, did not [18]. Guyton theorized that the IVC collapsed as it entered the thorax and that a collapsible vessel could not transmit a negative pressure upstream [18]. The first demonstration of this phenomenon in humans was demonstrated in a study on acute severe asthma [68].

In a healthy, spontaneously breathing subject, cyclic variations in the pleural pressure transmitted to the right atrial cavity lead to cyclical fluctuations in the RV preload. By increasing inspiration, the inspiratory IVC diameter may be reduced by 50 % [66]. When the IVC is expanded, the relationship between the diameter and the pressure is located at its horizontal portion. The respiratory variations in IVC diameter caused by a pressure drop in inspiration are abolished, for instance, in cardiac tamponade [69] or in severe RV failure [70]. In a patient subjected to positive pressure ventilation, the inspiratory phase leads to an increase in pleural pressure, which is transmitted to the cardiac fossa, thereby reducing the VR. The result is an inversion of the cyclical changes in the IVC diameter, increasing during inspiration and decreasing during expiration.

These respiratory changes are abolished by a dilatation of the vena cava, fact which corroborates a high RAP value (Fig. 6.7). The IVC respiratory variations in a patient under positive pressure mechanical ventilation are only observed when the RAP is normal or low. In a patient with circulatory failure, this finding can be observed as hypovolemia or preload dependency. Measuring the IVC diameter in mechanically ventilated patients does not accurately predict the RAP [71]. However, the absence of respiratory variations in the IVC diameter in mechanically ventilated patients with signs of circulatory failure suggests that volume expansion, if administered, will be ineffective in 90 % of cases [72].

Feissel et al. used respiratory variations in the IVC diameter to predict fluid responsiveness in mechanically ventilated patients [72]. The variability of the IVC diameter was calculated using the following formula:

$$\Delta IVC(\%) = \frac{D_{IVCmax} - D_{IVmin}}{\left(D_{IVCmax} + D_{IVCmin}\right)/2} \times 100$$

An increase in IVC diameter of 12 % during the inspiratory phase can differentiate responders from nonresponders with a positive predictive value of 93 % and a negative predictive value of 92 % [72]. This work shows that a noninvasive

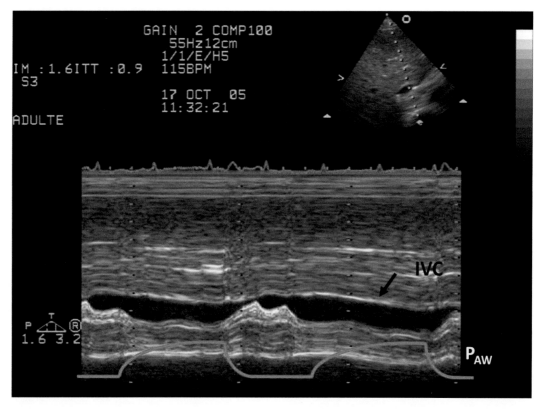

**Fig. 6.7** Subxyphoid transthoracic ultrasound centered on the inferior vena cava (IVC) in time-motion mode showing the IVC respiratory variation based on airway pressure ($P_{aw}$) in a patient with preload dependency who is under mechanical ventilation (Adapted from Feissel et al. [72]). Note that when the preload-dependent patient is placed under mechanical ventilation, IVC collapse occurs in expirium, whereas when the patient is breathing spontaneously, IVC collapse occurs in inspirium

dynamic parameter can assess the potential benefit of volume expansion. In addition, examination of the IVC is particularly easy and requires only limited experience in echocardiography. Other teams have confirmed these assumptions [73]. However, an unknown remains: is this index still valid in the case of high abdominal pressure wherein the variations in the IVC diameter are limited?

Another phenomenon previewed in mechanically ventilated patient is the presence of retrograde flow that results not from a tricuspid regurgitation but rather from cyclic compression of the right atrial wall by the lung. This compression of the RA pushes residual blood cells back toward the IVC, especially during cardiac systole, while the tricuspid valve is closed to prevent ante-

grade flow [73]. This retrograde flow is a major cause of the inaccuracy of the measurement by thermodilution in mechanically ventilated patients [74]. Anyway, inferior vena cava examination by echocardiography provides new accurate and easily acquired indices on the fluid responsiveness in mechanically ventilated patients and in patients with acute circulatory failure.

### 6.2.6   Measurements of Superior Vena Cava Respiratory Variations (ΔSVC)

Ultrasound examination of the SVC can be made by transesophageal echocardiography [26]. To keep this highly collapsible vessel

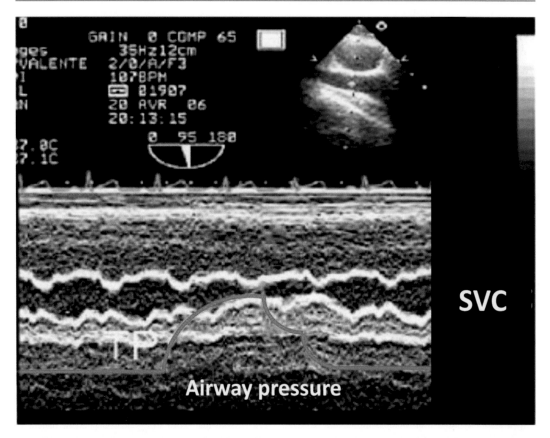

**Fig. 6.8** Ultrasound performed with a multiplane trans-esophageal probe oriented at 90° and centered on the superior vena cava (*SVC*) in time-motion mode in a preload-dependent, mechanically ventilated patient showing respiratory variations (Adapted from Vieillard-Baron et al. [26]). Collapse of the SVC in a preload-dependent patient occurs during the mechanical breath

open, it must be subjected to distension pressure, which should be higher than its critical closing pressure. In a patient placed under mechanical ventilation with positive pressure, the inspiratory phase leads to a lesser elevation in the RAP value than in the surrounding pressure (pleural pressure). Thus, the SVC, which is exposed to increased pleural pressure, behaves as a Starling "resistor." Therefore, the impact of a mechanical breath with the followed increase in intrathoracic pressure on SVC diameter is mainly determined by the amount of blood volume inside the present venous vessel.

An index of "collapsibility" of the SVC has been proposed and is calculated as

$$I_{\text{Coll}}(\%) = \frac{D_{\text{SVCmax}} - D_{\text{SVCmin}}}{S_{\text{SVCmax}}} \times 100 \quad [26]$$

This index allows for the prediction of fluid responsiveness in mechanically ventilated patients with positive pressure and acute circulatory failure [26]. This assessment requires examination of the SVC by a longitudinal section through the long axis using a multiplane trans-esophageal echocardiography probe (Fig. 6.8), recording of the two-dimensional image, and performance of a time-motion study. In one study, a collapsibility index of greater than 36 % was predictive, with a sensitivity of 90 % and a specificity of 100 %. A positive response to volume expansion was associated with a significant increase in the CO [26].

### 6.2.7  Measurement of the Pre-ejection Period and Ventricular Ejection Time

The determination of systolic time intervals was often used by cardiologists to monitor hypertrophic cardiomyopathy, ischemic heart disease, valvular heart disease, hypertensive disease, and pharmacological effects on the myocardium function [75]. The measurements were performed with phonocardiograhic recordings and carotidograms, Doppler echocardiograms or esophageal Doppler [76], and ECG recordings as a noninvasive method for assessing left ventricular function [77]. The pre-ejection period (PEP), the left ventricular ejection time (LVET), and the PEP/LVET ratio were the most commonly used systolic time intervals. In the past 20 years, as part of a noninvasive hemodynamic monitoring impetus in the ICU, these different times and phases of ventricular systole were also assessed using esophageal Doppler or impedance [76, 78].

The PEP is the time interval between ventricular depolarization (at the beginning of the QRS), beginning from the start of left ventricular ejection, characterized by opening of the aortic valve [79]. This time interval depends on the preload, afterload, and contractility of the left ventricle [80]. The normal range of the PEP in healthy subjects is about 100 ms [81], with several demonstrations that the heart rate does not appear to mainly influence the PEP value [80].

The LVET corresponds to the systolic time interval during which the left ventricle ejects blood into the aorta and is characterized by the opening-closing cycle of the aortic valve [80]. The time corresponding to the closing of the aortic valve (the aortic component of the second sound) is identified by phonocardiography or echocardiography. This time interval is essentially dependent on the stroke volume [82]. The normal range in healthy adults is $292 \pm 18$ ms [81]. This value is inversely proportional to the heart rate because the stroke volume decreases as the heart rate increases [82]. The PEP/ET ratio is a reliable index of left ventricular function [80]. Indeed, in the case of cardiac dysfunction, the PEP increases, and the ET decreases [80].

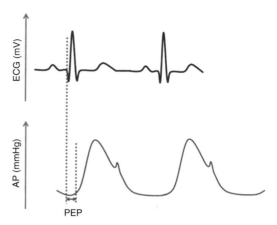

**Fig 6.9** Measurement technique of the pre-ejection period (*PEP*) from simultaneous recording of the electrocardiogram (*ECG*) and the invasive blood pressure (*AP*). The interval is measured from the beginning of the electrical ventricular depolarization (*Q*-wave) to the beginning of the rise in pressure of the AP

Therefore, the PEP/LVET ratio increases if there is a reduction in contractility. The normal range in healthy adults is $0.37 \pm 0.03$ [81]. There is a good correlation between this ratio and the various indices for assessing left ventricular performance [75, 83, 84].

In the ICU, all patients have continuous ECG monitors and are almost all equipped with an arterial catheter for the continuous measurement of blood pressure. It is then possible to simultaneously measure the PEP with the recording of the blood pressure waveform and the electrocardiogram [85]. The PEP is the time between the *Q*- or *R*-wave of the ECG and the start of the rise in blood pressure, as measured on the radial artery [85] (Fig. 6.9). Studies have shown that the PEP decreases when the preload increases [79] and that the PEP decreases when the SV increases [86]. In addition, various studies in humans have shown that the decrease in the PEP after volume expansion is associated with an increase in the SV [87] and that mechanical ventilation increases the PEP while decreasing the venous return and thus the left ventricular stroke volume [88]. The study of the respiratory variation of the PEP ($\Delta$PEP) in patients under mechanical ventilation has shown that when the patient is preload dependent, the PEP shortens during insufflation and increases during expiration [85]. The ΔPEP was calculated using the following formula:

$$\Delta PEP(\%) = \frac{PEP_{max} - PEP_{min}}{(PEP_{max} + PEP_{min})/2} \times 100$$

One study showed that the ΔPEP is a predictive dynamic index of fluid responsiveness [85]. Using the same principle but replacing the blood pressure signal with a plethysmography curve, a study showed that a ΔPEP >4 % is able to predict positive fluid responsiveness with a sensitivity of 100 % and a specificity of 67 % [89].

# References

1. Vincent JL, Weil MH (2006) Fluid challenge revisited. Crit Care Med 34(5):1333–1337
2. Michard F, Teboul JL (2000) Using heart-lung interactions to assess fluid responsiveness during mechanical ventilation. Crit Care 4(5):282–289
3. Michard F, Teboul JL (2002) Predicting fluid responsiveness in ICU patients: a critical analysis of the evidence. Chest 121(6):2000–2008
4. Osman D, Ridel C, Ray P, Monnet X, Anguel N, Richard C et al (2007) Cardiac filling pressures are not appropriate to predict hemodynamic response to volume challenge. Crit Care Med 35(1):64–68
5. Feissel M, Michard F, Mangin I, Ruyer O, Faller JP, Teboul JL (2001) Respiratory changes in aortic blood velocity as an indicator of fluid responsiveness in ventilated patients with septic shock. Chest 119(3):867–873
6. Tavernier B, Makhotine O, Lebuffe G, Dupont J, Scherpereel P (1998) Systolic pressure variation as a guide to fluid therapy in patients with sepsis-induced hypotension. Anesthesiology 89(6):1313–1321
7. Monnet X, Teboul JL (2006) Invasive measures of left ventricular preload. Curr Opin Crit Care 12(3):235–240
8. Teboul JL (2005) SRLF experts recommendations: indicators of volume resuscitation during circulatory failure. Ann Fr Anesth Reanim 24(5):568–576; quiz 77–81
9. Pinsky MR (1984) Determinants of pulmonary arterial flow variation during respiration. J Appl Physiol Respir Environ Exerc Physiol 56(5):1237–1245
10. Wallis TW, Robotham JL, Compean R, Kindred MK (1983) Mechanical heart-lung interaction with positive end-expiratory pressure. J Appl Physiol Respir Environ Exerc Physiol 54(4):1039–1047
11. Garcia X, Pinsky MR (2011) Clinical applicability of functional hemodynamic monitoring. Ann Intensive Care 1:35
12. Marik PE, Monnet X, Teboul JL (2011) Hemodynamic parameters to guide fluid therapy. Ann Intensive Care 1(1):1
13. Monnet X, Rienzo M, Osman D, Anguel N, Richard C, Pinsky MR et al (2006) Passive leg raising predicts fluid responsiveness in the critically ill. Crit Care Med 34(5):1402–1407
14. Reich DL, Konstadt SN, Raissi S, Hubbard M, Thys DM (1989) Trendelenburg position and passive leg raising do not significantly improve cardiopulmonary performance in the anesthetized patient with coronary artery disease. Crit Care Med 17(4):313–317
15. Thomas M, Shillingford J (1965) The circulatory response to a standard postural change in ischaemic heart disease. Br Heart J 27:17–27
16. Rocha P, Lemaigre D, Leroy M, Desfonds P, De Zuttere D, Liot F (1987) Nitroglycerin-induced decrease of carbon monoxide diffusion capacity in acute myocardial infarction reversed by elevating legs. Crit Care Med 15(2):131–133
17. Takagi S, Yokota M, Iwase M, Yoshida J, Hayashi H, Sotobata I et al (1989) The important role of left ventricular relaxation and left atrial pressure in the left ventricular filling velocity profile. Am Heart J 118(5 Pt 1):954–962
18. Guyton AC, Lindsey AW, Abernathy B, Richardson T (1957) Venous return at various right atrial pressures and the normal venous return curve. Am J Physiol 189(3):609–615
19. Chihara E, Hashimoto S, Kinoshita T, Hirose M, Tanaka Y, Morimoto T (1992) Elevated mean systemic filling pressure due to intermittent positive-pressure ventilation. Am J Physiol 262(4 Pt 2): H1116–H1121
20. Wong DH, Tremper KK, Zaccari J, Hajduczek J, Konchigeri HN, Hufstedler SM (1988) Acute cardiovascular response to passive leg raising. Crit Care Med 16(2):123–125
21. Boulain T, Achard JM, Teboul JL, Richard C, Perrotin D, Ginies G (2002) Changes in BP induced by passive leg raising predict response to fluid loading in critically ill patients. Chest 121(4):1245–1252
22. Lafanechere A, Pene F, Goulenok C, Delahaye A, Mallet V, Choukroun G et al (2006) Changes in aortic blood flow induced by passive leg raising predict fluid responsiveness in critically ill patients. Crit Care 10(5):R132
23. Biais M, Vidil L, Sarrabay P, Cottenceau V, Revel P, Sztark F (2009) Changes in stroke volume induced by passive leg raising in spontaneously breathing patients: comparison between echocardiography and Vigileo/FloTrac device. Crit Care 13(6):R195
24. Vieillard-Baron A, Chergui K, Augarde R, Prin S, Page B, Beauchet A et al (2003) Cyclic changes in arterial pulse during respiratory support revisited by Doppler echocardiography. Am J Respir Crit Care Med 168(6):671–676
25. Vieillard-Baron A, Augarde R, Prin S, Page B, Beauchet A, Jardin F (2001) Influence of superior vena caval zone condition on cyclic changes in right ventricular outflow during respiratory support. Anesthesiology 95(5):1083–1088 [Clinical Trial]

26. Vieillard-Baron A, Chergui K, Rabiller A, Peyrouset O, Page B, Beauchet A et al (2004) Superior vena caval collapsibility as a gauge of volume status in ventilated septic patients. Intensive Care Med 30(9):1734–1739 [Clinical Trial]

27. Perel A, Pizov R, Cotev S (1987) Systolic blood pressure variation is a sensitive indicator of hypovolemia in ventilated dogs subjected to graded hemorrhage. Anesthesiology 67(4):498–502

28. Perel A, Pizov R, Cotev S (2014) Respiratory variations in the arterial pressure during mechanical ventilation reflect volume status and fluid responsiveness. Intensive Care Med 40(6):798–807

29. Goedje O, Hoeke K, Lichtwarck-Aschoff M, Faltchauser A, Lamm P, Reichart B (1999) Continuous cardiac output by femoral arterial thermodilution calibrated pulse contour analysis: comparison with pulmonary arterial thermodilution. Crit Care Med 27(11):2407–2412 [Comparative Study]

30. Michard F, Boussat S, Chemla D, Anguel N, Mercat A, Lecarpentier Y et al (2000) Relation between respiratory changes in arterial pulse pressure and fluid responsiveness in septic patients with acute circulatory failure. Am J Respir Crit Care Med 162(1):134–138

31. Pizov R, Cohen M, Weiss Y, Segal E, Cotev S, Perel A (1996) Positive end-expiratory pressure-induced hemodynamic changes are reflected in the arterial pressure waveform. Crit Care Med 24(8):1381–1387

32. Pizov R, Ya'ari Y, Perel A (1988) Systolic pressure variation is greater during hemorrhage than during sodium nitroprusside-induced hypotension in ventilated dogs. Anesth Analg 67(2):170–174

33. Pizov R, Ya'ari Y, Perel A (1989) The arterial pressure waveform during acute ventricular failure and synchronized external chest compression. Anesth Analg 68(2):150–156

34. Szold A, Pizov R, Segal E, Perel A (1989) The effect of tidal volume and intravascular volume state on systolic pressure variation in ventilated dogs. Intensive Care Med 15(6):368–371

35. Coriat P, Vrillon M, Perel A, Baron JF, Le Bret F, Saada M et al (1994) A comparison of systolic blood pressure variations and echocardiographic estimates of end-diastolic left ventricular size in patients after aortic surgery. Anesth Analg 78(1):46–53

36. Perel A (1998) Assessing fluid responsiveness by the systolic pressure variation in mechanically ventilated patients. Systolic pressure variation as a guide to fluid therapy in patients with sepsis-induced hypotension. Anesthesiology 89(6):1309–1310

37. Stoneham MD (1999) Less is more … using systolic pressure variation to assess hypovolaemia. Br J Anaesth 83(4):550–551

38. Michard F, Chemla D, Richard C, Wysocki M, Pinsky MR, Lecarpentier Y et al (1999) Clinical use of respiratory changes in arterial pulse pressure to monitor the hemodynamic effects of PEEP. Am J Respir Crit Care Med 159(3):935–939

39. De Backer D, Taccone FS, Holsten R, Ibrahimi F, Vincent JL (2009) Influence of respiratory rate on stroke volume variation in mechanically ventilated patients. Anesthesiology 110(5):1092–1097

40. Kim HK, Pinsky MR (2008) Effect of tidal volume, sampling duration, and cardiac contractility on pulse pressure and stroke volume variation during positive-pressure ventilation. Crit Care Med 36(10):2858–2862 [Comparative Study Research Support, NIH, Extramural]

41. Muller L, Louart G, Bousquet PJ, Candela D, Zoric L, de La Coussaye JE et al (2010) The influence of the airway driving pressure on pulsed pressure variation as a predictor of fluid responsiveness. Intensive Care Med 36(3):496–503

42. Vistisen ST, Koefoed-Nielsen J, Larsson A (2010) Should dynamic parameters for prediction of fluid responsiveness be indexed to the tidal volume? Acta Anaesthesiol Scand 54(2):191–198 [Comparative Study Research Support, Non-US Gov't]

43. Duperret S, Lhuillier F, Piriou V, Vivier E, Metton O, Branche P et al (2007) Increased intra-abdominal pressure affects respiratory variations in arterial pressure in normovolaemic and hypovolaemic mechanically ventilated healthy pigs. Intensive Care Med 33(1):163–171

44. Huntsman LL, Stewart DK, Barnes SR, Franklin SB, Colocousis JS, Hessel EA (1983) Noninvasive Doppler determination of cardiac output in man. Clin Valid Circ 67(3):593–602

45. Slama M, Masson H, Teboul JL, Arnout ML, Susic D, Frohlich E et al (2002) Respiratory variations of aortic VTI: a new index of hypovolemia and fluid responsiveness. Am J Physiol Heart Circ Physiol 283(4):H1729–H1733

46. Monnet X, Rienzo M, Osman D, Anguel N, Richard C, Pinsky MR et al (2005) Esophageal Doppler monitoring predicts fluid responsiveness in critically ill ventilated patients. Intensive Care Med 31(9):1195–1201

47. Cannesson M, Tran NP, Cho M, Hatib F, Michard F (2012) Predicting fluid responsiveness with stroke volume variation despite multiple extrasystoles. Crit Care Med 40(1):193–198

48. Cannesson M, Besnard C, Durand PG, Bohe J, Jacques D (2005) Relation between respiratory variations in pulse oximetry plethysmographic waveform amplitude and arterial pulse pressure in ventilated patients. Crit Care 9(5):R562–R568

49. Feissel M, Teboul JL, Merlani P, Badie J, Faller JP, Bendjelid K (2007) Plethysmographic dynamic indices predict fluid responsiveness in septic ventilated patients. Intensive Care Med 33(6):993–999

50. Cannesson M, Desebbe O, Hachemi M, Jacques D, Bastien O, Lehot JJ (2007) Respiratory variations in pulse oximeter waveform amplitude are influenced by venous return in mechanically ventilated patients under general anaesthesia. Eur J Anaesthesiol 24(3):245–251 [Clinical Trial]

51. Monnet X, Lamia B, Teboul JL (2005) Pulse oximeter as a sensor of fluid responsiveness: do we have our finger on the best solution? Crit Care 9(5):429–430

52. Heenen S, De Backer D, Vincent JL (2006) How can the response to volume expansion in patients with spontaneous respiratory movements be predicted? Crit Care 10(4):R102

53. Delerme S, Castro S, Freund Y, Nazeyrollas P, Josse MO, Madonna-Py B et al (2010) Relation between pulse oximetry plethysmographic waveform amplitude induced by passive leg raising and cardiac index in spontaneously breathing subjects. Am J Emerg Med 28(4):505–510

54. Keller G, Cassar E, Desebbe O, Lehot JJ, Cannesson M (2008) Ability of pleth variability index to detect hemodynamic changes induced by passive leg raising in spontaneously breathing volunteers. Crit Care 12(2):R37

55. Cavallaro F, Sandroni C, Marano C, La Torre G, Mannocci A, De Waure C et al (2010) Diagnostic accuracy of passive leg raising for prediction of fluid responsiveness in adults: systematic review and meta-analysis of clinical studies. Intensive Care Med 36(9):1475–1483

56. Broch O, Bein B, Gruenewald M, Hocker J, Schottler J, Meybohm P et al (2011) Accuracy of the pleth variability index to predict fluid responsiveness depends on the perfusion index. Acta Anaesthesiol Scand 55(6):686–693

57. Yamaura K, Irita K, Kandabashi T, Tohyama K, Takahashi S (2007) Evaluation of finger and forehead pulse oximeters during mild hypothermic cardiopulmonary bypass. J Clin Monit Comput 21(4):249–252

58. Antonelli M, Levy M, Andrews PJ, Chastre J, Hudson LD, Manthous C et al (2007) Hemodynamic monitoring in shock and implications for management. International Consensus Conference, Paris, France, 27–28 April 2006. Intensive Care Med 33(4):575–590

59. Biais M, Cottenceau V, Petit L, Masson F, Cochard JF, Sztark F (2011) Impact of norepinephrine on the relationship between pleth variability index and pulse pressure variations in ICU adult patients. Crit Care 15(4):R168

60. Landsverk SA, Hoiseth LO, Kvandal P, Hisdal J, Skare O, Kirkeboen KA (2008) Poor agreement between respiratory variations in pulse oximetry photoplethysmographic waveform amplitude and pulse pressure in intensive care unit patients. Anesthesiology 109(5):849–855

61. Cannesson M, Awad AA, Shelley K (2009) Oscillations in the plethysmographic waveform amplitude: phenomenon hides behind artifacts. Anesthesiology 111(1):206–207; author reply 7–8

62. Desgranges FP, Desebbe O, Ghazouani A, Gilbert K, Keller G, Chiari P et al (2011) Influence of the site of measurement on the ability of plethysmographic variability index to predict fluid responsiveness. Br J Anaesth 107(3):329–335

63. Awad AA, Ghobashy MA, Ouda W, Stout RG, Silverman DG, Shelley KH (2001) Different responses of ear and finger pulse oximeter wave form to cold pressor test. Anesth Analg 92(6):1483–1486

64. Bendjelid K (2008) The pulse oximetry plethysmographic curve revisited. Curr Opin Crit Care 14(3):348–353

65. Cannesson M, Desebbe O, Rosamel P, Delannoy B, Robin J, Bastien O et al (2008) Pleth variability index to monitor the respiratory variations in the pulse oximeter plethysmographic waveform amplitude and predict fluid responsiveness in the operating theatre. Br J Anaesth 101(2):200–206

66. Mintz GS, Kotler MN, Parry WR, Iskandrian AS, Kane SA (1981) Reat-time inferior vena caval ultrasonography: normal and abnormal findings and its use in assessing right-heart function. Circulation 64(5):1018–1025

67. Nakao S, Come PC, McKay RG, Ransil BJ (1987) Effects of positional changes on inferior vena caval size and dynamics and correlations with right-sided cardiac pressure. Am J Cardiol 59(1):125–132 [Research Support, US Gov't, PHS]

68. Jardin F, Farcot JC, Boisante L, Prost JF, Gueret P, Bourdarias JP (1982) Mechanism of paradoxic pulse in bronchial asthma. Circulation 66(4):887–894 [Research Support, Non-US Gov't]

69. Himelman RB, Kircher B, Rockey DC, Schiller NB (1988) Inferior vena cava plethora with blunted respiratory response: a sensitive echocardiographic sign of cardiac tamponade. J Am Coll Cardiol 12(6):1470–1477 [Research Support, US Gov't, PHS]

70. Moreno FL, Hagan AD, Holmen JR, Pryor TA, Strickland RD, Castle CH (1984) Evaluation of size and dynamics of the inferior vena cava as an index of right-sided cardiac function. Am J Cardiol 53(4):579–585

71. Jue J, Chung W, Schiller NB (1992) Does inferior vena cava size predict right atrial pressures in patients receiving mechanical ventilation? J Am Soc Echocardiogr 5(6):613–619

72. Feissel M, Michard F, Faller JP, Teboul JL (2004) The respiratory variation in inferior vena cava diameter as a guide to fluid therapy. Intensive Care Med 30(9):1834–1837

73. Barbier C, Loubieres Y, Schmit C, Hayon J, Ricome JL, Jardin F et al (2004) Respiratory changes in inferior vena cava diameter are helpful in predicting fluid responsiveness in ventilated septic patients. Intensive Care Med 30(9):1740–1746

74. Jullien T, Valtier B, Hongnat JM, Dubourg O, Bourdarias JP, Jardin F (1995) Incidence of tricuspid regurgitation and vena caval backward flow in mechanically ventilated patients. A color Doppler and contrast echocardiographic study. Chest 107(2):488–493 [Research Support, Non-US Gov't]

75. Weissler AM (1987) The systolic time intervals and risk stratification after acute myocardial infarction. J Am Coll Cardiol 9(1):161–162

76. Singer M, Allen MJ, Webb AR, Bennett ED (1991) Effects of alterations in left ventricular filling, contractility, and systemic vascular resistance on the ascending aortic blood velocity waveform of normal subjects. Crit Care Med 19(9):1138–1145

77. Weissler AM, Harris WS, Schoenfeld CD (1969) Bedside technics for the evaluation of ventricular function in man. Am J Cardiol 23(4):577–583

78. Tournadre JP, Muchada R, Lansiaux S, Chassard D (1999) Measurements of systolic time intervals using a transoesophageal pulsed echo-Doppler. Br J Anaesth 83(4):630–636

79. Weissler AM (1977) Current concepts in cardiology. Systolic-time intervals. N Engl J Med 296(6): 321–324

80. Boudoulas H (1990) Systolic time intervals. Eur Heart J 11(Suppl I):93–104 [Review]

81. Mertens HM, Mannebach H, Trieb G, Gleichmann U (1981) Influence of heart rate on systolic time intervals: effects of atrial pacing versus dynamic exercise. Clin Cardiol 4(1):22–27

82. Ferro G, Ricciardelli B, Sacca L, Chiariello M, Volpe M, Tari MG et al (1980) Relationship between systolic time intervals and heart rate during atrial or ventricular pacing in normal subjects. Jpn Heart J 21(6):765–771 [Research Support, Non-US Gov't]

83. Hamada M, Ito T, Hiwada K, Kokubu T, Genda A, Takeda R (1991) Characteristics of systolic time intervals in patients with pheochromocytoma. Jpn Circ J 55(5):417–426

84. Shoemaker WC, Wo CC, Bishop MH, Appel PL, Van de Water JM, Harrington GR et al (1994) Multicenter trial of a new thoracic electrical bioimpedance device for cardiac output estimation. Crit Care Med 22(12):1907–1912 [Clinical Trial Comparative Study Multicenter Study Research Support, US Gov't, PHS]

85. Bendjelid K, Suter PM, Romand JA (2004) The respiratory change in preejection period: a new method to predict fluid responsiveness. J Appl Physiol 96(1):337–342

86. Wallace AG, Mitchell JH, Skinner NS, Sarnoff SJ (1963) Duration of the phases of left ventricular systole. Circ Res 12:611–619

87. Matsuno Y, Morioka S, Murakami Y, Kobayashi S, Moriyama K (1988) Mechanism of prolongation of pre-ejection period in the hypertrophied left ventricle with normal systolic function in unanesthetized hypertensive dogs. Clin Cardiol 11(10):702–706

88. Brundin T, Hedenstierna G, McCarthy G (1976) Effect of intermittent positive pressure ventilation on cardiac systolic time intervals. Acta Anaesthesiol Scand 20(4):278–284

89. Feissel M, Badie J, Merlani PG, Faller JP, Bendjelid K (2005) Pre-ejection period variations predict the fluid responsiveness of septic ventilated patients. Crit Care Med 33(11):2534–2539

# Perspectives

Hemodynamic monitoring in the ICU has continued to evolve over the years. On the one hand, the industry constantly offers new tools that can be methodologically validated to provide clinicians with the terms of use and limitations of these devices. On the other hand, researchers are attempting to innovate and discover new indices and/or markers that can replace discontinuous and invasive measures of cardiac output and indices that can predict fluid responsiveness in patients who are in a state of shock. In the present book, the authors revealed various hemodynamic monitoring techniques available for the intensivist. As we have shown, each technique has its advantages and limitations. For several years, the subsidized intensive care industry has developed monitoring devices that measure various parameters and are less invasive than the classical pulmonary artery catheter. Nevertheless, the cardiac output remains the most significant and important hemodynamic variable and marker.

The abilities of various hemodynamic monitoring devices to measure hemodynamic parameters were reviewed. The deduction is clearly exposed as the techniques that allow for the reliable measurement of cardiac output have been and remain the most invasive techniques: pulmonary artery catheterization and transpulmonary thermodilution. Transpulmonary thermodilution appears to have increased in popularity over the pulmonary artery catheter. Nonetheless, the PAC is still used, especially in academic centers. On the other hand, echocardiography remains a noninvasive technique that is both accurate and reproducible. Indeed, it measures the cardiac output, estimates preload, and anatomically evaluates heart-vessel structures and pericardial constraint; no other technique was able to match these properties. However, there are two major issues with this technique: (1) it requires an advanced learning curve; (2) it does not allow for the continuous monitoring of patients and is difficult to implement in intensive care units [1, 2].

The introduction of new, less invasive technologies providing the ability to simplify the decision may appear attractive. In addition to the stroke volume and cardiac output, hemodynamic monitoring devices provide various additional hemodynamic variables (Table 7.1), including static preload variables, functional hemodynamic variables, and the continuous central venous oxygen saturation ($ScvO_2$). However, these noninvasive methods of cardiac output monitoring should not be naively used. Intensivists must ensure that proper validation studies were conducted and involved the right categories of patients. For example, the reliability of these minimally invasive devices for continuous monitoring of the cardiac output is good in patients not treated with catecholamines [3]. Conversely, these devices do not meet the standards of accuracy when patients receive high doses of vasopressors [4, 5]. Users must be aware of the inherent limitations of each device. Therefore, the most suitable device should be used to collect the required data for determining the appropriate therapeutic goal [6].

© Springer International Publishing Switzerland 2016
R. Giraud, K. Bendjelid, *Hemodynamic Monitoring in the ICU*, DOI 10.1007/978-3-319-29430-8_7

**Table 7.1** Available cardiac output monitoring systems with their respective advantages and disadvantages

| Technology | System | Invasiveness | Mechanism | Advantages | Disadvantages | Outcome studies |
|---|---|---|---|---|---|---|
| Pulmonary artery catheter | Vigilance | +++ | Thermodilution | Gold standard for continuous/intermittent cardiac output monitoring. Allows measurement of pulmonary pressures and mixed venous oxygen saturation | No dynamic parameters of fluid responsiveness. Provides cardiac output information every few minutes | − |
| Calibrated pulse contour analysis | PiCCO$_2$ | ++ | Transpulmonary thermodilution + pulse contour analysis | Continuous cardiac output monitoring. Central venous oxygen saturation with specific device. Good accuracy | Remains significantly invasive. Requires a specific femoral artery catheter | 0 |
| | VolumeView | + | Transpulmonary thermodilution + pulse contour analysis | Continuous cardiac output monitoring. Central venous oxygen saturation with specific device. Good accuracy | Remains significantly invasive. Requires a specific femoral artery catheter | 0 |
| | LiDCOplus | + | Lithium dilution | Continuous cardiac output monitoring | Lithium very expensive | + |
| Uncalibrated pulse contour analysis | PulsioFlex | + | Pulse wave analysis | Continuous cardiac output monitoring. Mini-invasive, self-calibrated systems. Can be used with any arterial line and arterial pressure sensor | No validation studies | 0 |
| | LiDCOrapid | + | Pulse wave analysis | Continuous cardiac output monitoring. Mini-invasive, self-calibrated systems. Can be used with any arterial line and arterial pressure sensor | Not enough validation studies | 0 |
| | FloTrac | + | Pulse wave analysis | Continuous cardiac output monitoring. Mini-invasive, self-calibrated systems | Accuracy of cardiac output has been a concern. Sensitive to changes in vasomotor tone. Requires a specific arterial pressure sensor | + |
| | PRAM | + | Pulse wave analysis | Continuous cardiac output monitoring. Mini-invasive, self-calibrated systems | Not enough validation studies. Requires a specific arterial kit | 0 |
| | Nexfin | 0 | Noninvasive pulse wave analysis | Continuous cardiac output monitoring. Completely noninvasive, self-calibrated system | Not enough validation study. Motion artifact | 0 |

| Ultrasound | Cardio Q | 0+ | Doppler ultrasound | Less invasive then arterial-based systems, qualifies for billable monitoring in the USA | Requires frequent manipulation for proper position, significant potential for user variability | +++ |
|---|---|---|---|---|---|---|
| | USCOM | 0 | Suprasternal ultrasound | Noninvasive cardiac output measurement | Intermittent. Operator dependent | 0 |
| Bioreactance | NiCOM | 0 | Bioreactance | Noninvasive continuous cardiac output monitoring | Few validation studies. Requires a specific arterial kit and a specific endotracheal tube | 0 |
| Endotracheal bioimpedance | ECOM | + | Bioimpedance | Mini-invasive and continuous cardiac output monitoring | Few validation studies. Requires a specific arterial kit and a specific endotracheal tube | 0 |
| Thoracic bioimpedance | BioZ | 0 | Bioimpedance | Noninvasive cardiac output measurement | Many negative studies in the critical care setting | 0 |

0, None; 0+, very slight; +, slight; ++, intermediate; +++, severe. PiCCO plus, Pulsion Medical Systems, Irving, TX, USA; VolumeView, Edwards, Irvine, CA, USA; LiDCOplus, LiDCO Ltd, London, UK; LiDCOrapid, LiDCO Ltd, London, UK; PulsioFlex, Pulsion Medical Systems, Irving, TX, USA; PRAM, Multiple Suppliers; Nexfin, BMEye, Amsterdam, Netherlands; Cardio Q, Deltex Medical Limited, Chichester, West Sussex, UK; USCOM, Uscom, Sydney, Australia; NiCOM, Cheetah Medical, Tel Aviv, Israel; ECOM, ConMed, Irvine, CA, USA; BioZ, CardioDynamics, San Diego, CA, USA

The current trend is to use less invasive monitoring devices [7]. However, although these tools have simplified hemodynamic calculations, they remain subject to restrictions and may lead to false results and false treatment decisions if not used properly. Several issues must be considered when introducing new technologies in clinical practice. First, the implementation of a new device typically requires clinical validation by comparing the new technology with a "gold standard" in a standardized clinical setting. Unfortunately, there is no "gold standard" for cardiac output measurements. Normally, pulmonary artery thermodilution (the "ice water bolus" technique) is considered the "clinical reference" and is used in most studies as a reference technique. However, this technique has limitations that can lead to erroneous results. In addition, standardized validation that is dependent on the patient's clinical condition, a defined number of measurements, and the induction of heart rate changes is rarely performed.

Second, the Bland-Altman analysis became the standard statistical method for comparing the cardiac output measurements of a new device with those of a reference technique [8]; however, it includes biases and concordance limits that are not always easy to interpret. Percentage of error calculation was recently introduced by Lester Critchley and is now required in most validation studies [9–11]. A percentage of error threshold of 30 % was initially defined as the criterion of acceptability for cardiac output measurements and was subsequently used as a reference value. This threshold, however, has recently been questioned on the basis of a large meta-analysis of all available minimally invasive measurement techniques [10]. There is now debate regarding the percentage of error that indicates acceptable clinical reliability. Therefore, in the future, clinicians and researchers must have access to new perspectives on the statistical procedures used to validate new techniques, such as the recently proposed "polar plot" concordance analysis [12]. Indeed, concordance allows only a rough estimate of trends, i.e., the percentage of cardiac output changes in the same direction measured by the two techniques; however, the polar plot analysis is a more accurate method that allows the

**Table 7.2** Systems allowing for the monitoring of dynamic parameters of fluid responsiveness

| Dynamic parameter of fluid responsiveness | Monitor available for their display |
|---|---|
| Systolic pressure variation | Can be eyeballed accurately |
| Pulse pressure variation | Cannot be eyeballed |
| | Philips IntelliVue monitors |
| | LiDCOrapid |
| | LiDCOplus |
| | PiCCO$_2$ |
| | PulsioFlex |
| | PRAM |
| | Nexfin |
| | CNAP |
| | General Electric Monitors |
| Stroke volume variation | LiDCOrapid |
| | LiDCOplus |
| | PiCCO plus |
| | PulsioFlex |
| | PRAM |
| | Vigileo FloTrac |
| | EV1000 VolumeView |
| | ECOM |
| | BioZ |
| | NICOM |
| Pleth variability index | Masimo Radical-7 |
| Passive leg raising | Demonstrated with esophageal Doppler, PiCCO$_2$, echocardiography, NICOM, and Vigileo FloTrac |
| Pre-ejection period | Not available in clinical practice |

quantification of trends in a manner analogous to that of the Bland-Altman analysis.

Considering the technical features and the typical limitations of the different cardiac output monitoring techniques, it is obvious that no single device can comply with all of the clinical requirements. The different hemodynamic monitoring tools available on the market have their advantages and disadvantages (Tables 7.1 and 7.2). Therefore, different devices may be used in an integrative manner along a typical clinical patient pathway based on the invasiveness of the devices and the availability of additional hemodynamic variables. In the presence of factors that affect the accuracy of all minimally invasive cardiac output monitoring devices or when pulmonary artery pressure monitoring or right heart failure treatment is required, PAC insertion may be required for patient-specific therapy.

Finally, our bedside experience suggests that no ideal hemodynamic monitoring device exists. To ensure the optimal management of hemodynamic parameters, different devices may be required to meet the needs of different patient groups and different clinical scenarios. Many currently available devices comply with these requirements; therefore, some minimally invasive devices such as additional monitoring tools should be distinguished. However, when the minimally invasive methods of assessing cardiac output have significant limitations or when continuous monitoring of pulmonary artery pressure is required, the integrative use of a pulmonary artery catheter should always be considered [13]. Only the correct use of a device, adequate hemodynamic management, and/or a protocol based on therapeutic goals may be able to reduce morbidity and mortality [14]. Indeed, regardless of the precision and accuracy of a hemodynamic monitoring device, its impact on the prognosis of patients with hemodynamic instability is entirely dependent on the decisions that are made once the measured values are obtained. Therefore, although a measurement device can offer considerable decision-making support, its effects on patient management are dependent on the intensivist.

# References

1. Giraud R, Siegenthaler N, Tagan D, Bendjelid K (2009) Evaluation of practical skills in echocardiography for intensivists. Rev Med Suisse 5(229):2518–2521
2. Giraud R, Siegenthaler N, Tagan D, Bendjelid K (2011) Evaluation of skills required to practice advanced echocardiography in intensive care. Rev Med Suisse 7(282):413–416
3. Tsai YF, Liu FC, Yu HP (2013) FloTrac/Vigileo system monitoring in acute-care surgery: current and future trends. Expert Rev Med Devices 10(6):717–728, Review
4. Metzelder S, Coburn M, Fries M, Reinges M, Reich S, Rossaint R et al (2011) Performance of cardiac output measurement derived from arterial pressure waveform analysis in patients requiring high-dose vasopressor therapy. Br J Anaesth 106(6):776–784 [Clinical Trial]
5. Suehiro K, Tanaka K, Funao T, Matsuura T, Mori T, Nishikawa K (2013) Systemic vascular resistance has an impact on the reliability of the Vigileo-FloTrac system in measuring cardiac output and tracking cardiac output changes. Br J Anaesth 111(2):170–177, Research Support, Non-U.S. Gov't
6. Arulkumaran N, Corredor C, Hamilton MA, Ball J, Grounds RM, Rhodes A et al (2014) Cardiac complications associated with goal-directed therapy in high-risk surgical patients: a meta-analysis. Br J Anaesth 112(4):648–659 [Meta-Analysis]
7. Thiele RH, Bartels K, Gan TJ (2015) Cardiac output monitoring: a contemporary assessment and review. Crit Care Med 43(1):177–185
8. Bland JM, Altman DG (2012) Agreed statistics: measurement method comparison. Anesthesiology 116(1):182–185
9. Critchley LA, Critchley JA (1999) A meta-analysis of studies using bias and precision statistics to compare cardiac output measurement techniques. J Clin Monit Comput 15(2):85–91, Comparative Study Meta-Analysis
10. Peyton PJ, Chong SW (2010) Minimally invasive measurement of cardiac output during surgery and critical care: a meta-analysis of accuracy and precision. Anesthesiology 113(5):1220–1235
11. Critchley LA (2011) Bias and precision statistics: should we still adhere to the 30% benchmark for cardiac output monitor validation studies? Anesthesiology 114(5):1245; author reply -6
12. Critchley LA, Lee A, Ho AM (2010) A critical review of the ability of continuous cardiac output monitors to measure trends in cardiac output. Anesth Analg 111(5):1180–1192
13. Vincent JL (2011) So we use less pulmonary artery catheters – but why? Crit Care Med 39(7):1820–1822
14. Vincent JL, Rhodes A, Perel A, Martin GS, Della Rocca G, Vallet B et al (2011) Clinical review: update on hemodynamic monitoring – a consensus of 16. Crit Care 15(4):229

Printed in the United States
By Bookmasters